ALSO BY LEONARD MLODINOW

A Briefer History of Time (with Stephen Hawking)

Feynman's Rainbow: A Search for Beauty in Physics and in Life

Euclid's Window: The Story of Geometry from Parallel Lines to Hyperspace

For children (with Matt Costello)

The Last Dinosaur

Titanic Cat

The Drunkard's Walk

The Drunkard's Walk

How Randomness Rules Our Lives

LEONARD MLODINOW

YORK

All rights reserved. Published in the United States by Pantheon Books,
a division of Random House, Inc., New York, and in Canada by
Random House of Canada Limited, Toronto.

Pantheon Books and colophon are registered trademarks of Random House, Inc.

Library of Congress Cataloging-in-Publication Data

Mlodinow, Leonard, [date]
The Drunkard's walk : how randomness rules our lives / Leonard Mlodinow.
p. cm.
Includes index.
ISBN 978-0-375-42404-5
1. Random variables.   2. Probabilities.   3. Chance.   I. Title.
QA273.M63 2008
519.2—dc22   2007042507

www.pantheonbooks.com

Printed in the United States of America

6 8 9 7

To my three miracles of randomness:
Olivia, Nicolai, and Alexei . . .
and for Sabina Jakubowicz

CONTENTS

Contents

PROLOGUE

A FEW YEARS AGO a man won the Spanish national lottery with a ticket that ended in the number 48. Proud of his "accomplishment," he revealed the theory that brought him the riches. "I dreamed of the number 7 for seven straight nights," he said, "and 7 times 7 is 48."[1] Those of us with a better command of our multiplication tables might chuckle at the man's error, but we all create our own view of the world and then employ it to filter and process our perceptions, extracting meaning from the ocean of data that washes over us in daily life. And we often make errors that, though less obvious, are just as significant as his.

The fact that human intuition is ill suited to situations involving uncertainty was known as early as the 1930s, when researchers noted that people could neither make up a sequence of numbers that passed mathematical tests for randomness nor recognize reliably whether a given string was randomly generated. In the past few decades a new academic field has emerged to study how people make judgments and decisions when faced with imperfect or incomplete information. Their research has shown that when chance is involved, people's thought processes are often seriously flawed. The work draws from many disciplines, from mathematics and the traditional sciences as well as cognitive psychology, behavioral economics, and modern neuroscience. But although such studies were legitimated by a recent Nobel Prize (in Economics), their lessons for the most part have not trickled down from academic circles to the popular psyche. This book is an attempt to remedy that. It is about the principles that govern chance, the development of those ideas, and the manner in which they play out in politics, business, medicine, economics, sports, leisure, and other areas of human affairs. It is also about the way we make choices and the processes that lead us to make mistaken judgments and poor decisions when confronted with randomness or uncertainty.

Information that is lacking often invites competing interpretations. That's why such great effort was required to confirm global warming, why drugs are sometimes declared safe and then pulled from the market, and presumably why not everyone agrees with my observation that chocolate milkshakes are an indispensable component of a heart-healthy diet. Unfortunately the misinterpretation of data has many negative consequences, both large and small. As we'll see, for example, both doctors and patients often misinterpret statistics regarding the effectiveness of drugs and the meaning of important medical tests. Parents, teachers, and students misunderstand the significance of exams such as the SAT, and wine connoisseurs make the same mistakes about wine ratings. Investors draw invalid conclusions from the historical performance of mutual funds.

In sports we have developed a culture in which, based on intuitive feelings of correlation, a team's success or failure is often attributed largely to the ability of the coach. As a result, when teams fail, the coach is often fired. Mathematical analysis of firings in all major sports, however, has shown that those firings had, on average, no effect on team performance.[2] An analogous phenomenon occurs in the corporate world, where CEOs are thought to have superhuman power to make or break a company. Yet time and time again at Kodak, Lucent, Xerox, and other companies, that power has proved illusory. In the 1990s, for instance, when he ran GE Capital Services under Jack Welch, Gary Wendt was thought of as one of the smartest businessmen in the country. Wendt parlayed that reputation into a $45 million bonus when he was hired to run the troubled finance company Conseco. Investors apparently agreed that with Wendt at the helm, Conseco's troubles were over: the company's stock tripled within a year. But two years after that Wendt abruptly resigned, Conseco went bankrupt, and the stock was trading for pennies.[3] Had Wendt's task been impossible? Was he asleep at the wheel? Or had his coronation rested on questionable assumptions—for example, that an executive has a near-absolute ability to affect a company or a person's single past success is a reliable indicator of future performance? On any specific occasion one cannot be confident of the answers without examining the details of the situation at hand. I will do that in several instances in this book, but more important, I will present the tools needed to identify the footprints of chance.

To swim against the current of human intuition is a difficult task. As we'll see, the human mind is built to identify for each event a definite cause and can therefore have a hard time accepting the influence of unrelated or random factors. And so the first step is to realize that success or failure sometimes arises neither from great skill nor from great incompetence but from, as the economist Armen Alchian wrote, "fortuitous circumstances."[4] Random processes are fundamental in nature and are ubiquitous in our everyday lives, yet most people do not understand them or think much about them.

The title *The Drunkard's Walk* comes from a mathematical term describing random motion, such as the paths molecules follow as they fly through space, incessantly bumping, and being bumped by, their sister molecules. That can be a metaphor for our lives, our paths from college to career, from single life to family life, from first hole of golf to eighteenth. The surprise is that the tools used to understand the drunkard's walk can also be employed to help understand the events of everyday life. The goal of this book is to illustrate the role of chance in the world around us and to show how we may recognize it at work in human affairs. I hope that after this tour of the world of randomness, you, the reader, will begin to see life in a different light, with a deeper understanding of the everyday world.

The Drunkard's Walk

CHAPTER 1

Peering through the Eyepiece of Randomness

I REMEMBER, as a teenager, watching the yellow flame of the Sabbath candles dancing randomly above the white paraffin cylinders that fueled them. I was too young to think candlelight romantic, but still I found it magical—because of the flickering images created by the fire. They shifted and morphed, grew and waned, all without apparent cause or plan. Surely, I believed, there must be rhyme and reason underlying the flame, some pattern that scientists could predict and explain with their mathematical equations. "Life isn't like that," my father told me. "Sometimes things happen that cannot be foreseen." He told me of the time when, in Buchenwald, the Nazi concentration camp in which he was imprisoned and starving, he stole a loaf of bread from the bakery. The baker had the Gestapo gather everyone who might have committed the crime and line the suspects up. "Who stole the bread?" the baker asked. When no one answered, he told the guards to shoot the suspects one by one until either they were all dead or someone confessed. My father stepped forward to spare the others. He did not try to paint himself in a heroic light but told me that he did it because he expected to be shot either way. Instead of having him killed, though, the baker gave my father a plum job, as his assistant. "A chance event," my father said. "It had nothing to do with you, but had it happened differently, you would never have been born." It struck me then that I have Hitler to thank for my existence, for the Germans had killed my father's wife and two young children, erasing his prior life. And so were it not for the war, my father would never have emigrated to New York, never have met my mother, also a refugee, and never have produced me and my two brothers.

My father rarely spoke of the war. I didn't realize it then, but years later it dawned on me that whenever he shared his ordeals, it was not so much because he wanted me to know of his experiences but rather because he wanted to impart a larger lesson about life. War is an extreme circumstance, but the role of chance in our lives is not predicated on extremes. The outline of our lives, like the candle's flame, is continuously coaxed in new directions by a variety of random events that, along with our responses to them, determine our fate. As a result, life is both hard to predict and hard to interpret. Just as, looking at a Rorschach blot, you might see Madonna and I, a duck-billed platypus, the data we encounter in business, law, medicine, sports, the media, or your child's third-grade report card can be read in many ways. Yet interpreting the role of chance in an event is not like intepreting a Rorschach blot; there are right ways and wrong ways to do it.

We often employ intuitive processes when we make assessments and choices in uncertain situations. Those processes no doubt carried an evolutionary advantage when we had to decide whether a saber-toothed tiger was smiling because it was fat and happy or because it was famished and saw us as its next meal. But the modern world has a different balance, and today those intuitive processes come with drawbacks. When we use our habitual ways of thinking to deal with today's tigers, we can be led to decisions that are less than optimal or even incongruous. That conclusion comes as no surprise to those who study how the brain processes uncertainty: many studies point to a close connection between the parts of our brain that make assessments of chance situations and those that handle the human characteristic that is often considered our prime source of irrationality—our emotions. Functional magnetic resonance imaging, for example, shows that risk and reward are assessed by parts of the dopaminergic system, a brain-reward circuit important for motivational and emotional processes.[1] The images show, too, that the amygdala, which is also linked to our emotional state, especially fear, is activated when we make decisions couched in uncertainty.[2]

The mechanisms by which people analyze situations involving chance are an intricate product of evolutionary factors, brain structure, personal experience, knowledge, and emotion. In fact, the human response to uncertainty is so complex that sometimes different structures within the brain come to different conclusions and apparently fight it out to determine which one will dominate. For example, if your face swells to five times its normal size three out of every four times you eat shrimp, the "logical" left hemisphere of your brain will attempt to find a pattern. The "intuitive" right hemisphere of your brain, on the other hand, will simply say "avoid shrimp." At least that's what researchers found in less painful experimental setups. The game is called probability guessing. In lieu of toying with shrimp and histamine, subjects are shown a series of cards or lights, which can have two colors, say green and red. Things are arranged so that the colors will appear with different probabilities but otherwise without a pattern. For example, red might appear twice as often as green in a sequence like red-red-green-red-green-red-red-green-green-red-red-red, and so on. The task of the subject, after watching for a while, is to predict whether each new member of the sequence will be red or green.

The game has two basic strategies. One is to always guess the color that you notice occurs more frequently. That is the route favored by rats and other nonhuman animals. If you employ this strategy, you are guaranteed a certain degree of success but you are also conceding that you will do no better. For instance, if green shows up 75 percent of the time and you decide to always guess green, you will be correct 75 percent of the time. The other strategy is to "match" your proportion of green and red guesses to the proportion of green and red you observed in the past. If the greens and reds appear in a pattern and you can figure out the pattern, this strategy enables you to guess right every time. But if the colors appear at random, you would be better off sticking with the first strategy. In the case where green randomly appears 75 percent of the time, the second strategy will lead to the correct guess only about 6 times in 10.

Humans usually try to guess the pattern, and in the process we allow ourselves to be outperformed by a rat. But there are people with certain types of post-surgical brain impairment—called a split brain—that precludes the right and left hemispheres of the brain from communicating with each other. If the probability experiment is performed on these patients such that they see the colored light or card with only their left eye and employ only their left hand to signal their predictions, it amounts to an experiment on the right side of the brain. But if the experiment is performed so as to involve only their right eye and right hand, it is an experiment on the left brain. When researchers performed those experiments, they found that—in the same patients—the right hemisphere always chose to guess the more frequent color and the left hemisphere always tried to guess the pattern.[3]

Making wise assessments and choices in the face of uncertainty is a rare skill. But like any skill, it can be improved with experience. In the pages that follow, I will examine the role of chance in the world around us, the ideas that have been developed over the centuries to help us understand that role, and the factors that often lead us astray. The British philosopher and mathematician Bertrand Russell wrote,

> We all start from "naive realism," i.e., the doctrine that things are what they seem. We think that grass is green, that stones are hard, and that snow is cold. But physics assures us that the greenness of grass, the hardness of stones, and the coldness of snow are not the greenness of grass, the hardness of stones, and the coldness of snow that we know in our own experience, but something very different.[4]

In what follows we will peer at life through the eyepiece of randomness and see that many of the events of our lives, too, are not quite what they seem but rather something very different.

IN 2002 THE NOBEL COMMITTEE awarded the Nobel Prize in Economics to a scientist named Daniel Kahneman. Economists do all sorts of things these days—they explain why teachers are paid so little, why football teams are worth so much, and why bodily functions help set a limit on the size of hog farms (a hog excretes three to five times as much as a human, so a farm with thousands of hogs on it often produces more waste than the neighboring cities).[5] Despite all the great research generated by economists, the 2002 Nobel Prize was notable because Kahneman is not an economist. He is a psychologist, and for decades, with the late Amos Tversky, Kahneman studied and clarified the kinds of misperceptions of randomness that fuel many of the common fallacies I will talk about in this book.

The greatest challenge in understanding the role of randomness in life is that although the basic principles of randomness arise from everyday logic, many of the consequences that follow from those principles prove counterintuitive. Kahneman and Tversky's studies were themselves spurred by a random event. In the mid-1960s, Kahneman, then a junior psychology professor at Hebrew University, agreed to perform a rather unexciting chore: lecturing to a group of Israeli air force flight instructors on the conventional wisdom of behavior modification and its application to the psychology of flight training. Kahneman drove home the point that rewarding positive behavior works but punishing mistakes does not. One of his students interrupted, voicing an opinion that would lead Kahneman to an epiphany and guide his research for decades.[6]

"I've often praised people warmly for beautifully executed maneuvers, and the next time they always do worse," the flight instructor said. "And I've screamed at people for badly executed maneuvers, and by and large the next time they improve. Don't tell me that reward works and punishment doesn't work. My experience contradicts it." The other flight instructors agreed. To Kahneman the flight instructors' experiences rang true. On the other hand, Kahneman believed in the animal experiments that demonstrated that reward works better than punishment. He ruminated on this apparent paradox. And then it struck him: the screaming preceded the improvement, but contrary to appearances it did not cause it.

How can that be? The answer lies in a phenomenon called regression toward the mean. That is, in any series of random events an extraordinary event is most likely to be followed, due purely to chance, by a more ordinary one. Here is how it works: The student pilots all had a certain personal ability to fly fighter planes. Raising their skill level involved many factors and required extensive practice, so although their skill was slowly improving through flight training, the change wouldn't be noticeable from one maneuver to the next. Any especially good or especially poor performance was thus mostly a matter of luck. So if a pilot made an exceptionally good landing—one far above his normal level of performance—then the odds would be good that he would perform closer to his norm—that is, worse—the next day. And if his instructor had praised him, it would appear that the praise had done no good. But if a pilot made an exceptionally bad landing—running the plane off the end of the runway and into the vat of corn chowder in the base cafeteria—then the odds would be good that the next day he would perform closer to his norm—that is, better. And if his instructor had a habit of screaming "you clumsy ape" when a student performed poorly, it would appear that his criticism did some good. In this way an *apparent* pattern would emerge: student performs well, praise does no good; student performs poorly, instructor compares student to lower primate at high volume, student improves. The instructors in Kahneman's class had concluded from such experiences that their screaming was a powerful educational tool. In reality it made no difference at all.

This error in intuition spurred Kahneman's thinking. He wondered, are such misconceptions universal? Do we, like the flight instructors, believe that harsh criticism improves our children's behavior or our employees' performance? Do we make other misjudgments when faced with uncertainty? Kahneman knew that human beings, by necessity, employ certain strategies to reduce the complexity of tasks of judgment and that intuition about probabilities plays an important part in that process. Will you feel sick after eating that luscious-looking seviche tostada from the street vendor? You don't consciously recall all the comparable food stands you've patronized, count the number of times you've spent the following night guzzling Pepto-Bismol, and come up with a numerical estimate. You let your intuition do the work. But research in the 1950s and early '60s indicated that people's intuition about randomness fails them in such situations. How widespread, Kahneman wondered, was this misunderstanding of uncertainty? And what are its implications for human decision making? A few years passed, and Kahneman invited a fellow junior professor, Amos Tversky, to give a guest lecture at one of his seminars. Later, at lunch, Kahneman mentioned his developing ideas to Tversky. Over the next thirty years, Tversky and Kahneman found that even among sophisticated subjects, when it came to random processes—whether in military or sports situations, business quandaries, or medical questions— people's beliefs and intuition very often let them down.

Suppose four publishers have rejected the manuscript for your thriller about love, war, and global warming. Your intuition and the bad feeling in the pit of your stomach might say that the rejections by all those publishing experts mean your manuscript is no good. But is your intuition correct? Is your novel unsellable? We all know from experience that if several tosses of a coin come up heads, it doesn't mean we are tossing a two-headed coin. Could it be that publishing success is so unpredictable that even if our novel is destined for the best-seller list, numerous publishers could miss the point and send those letters that say thanks but no thanks? One book in the 1950s was rejected by publishers, who responded with such comments as "very dull," "a dreary record of typical family bickering, petty annoyances and adolescent emotions," and "even if the work had come to light five years ago, when the subject [World War II] was timely, I

9

don't see that there would have been a chance for it." That book, *The Diary of a Young Girl* by Anne Frank, has sold 30 million copies, making it one of the best-selling books in history. Rejection letters were also sent to Sylvia Plath because "there certainly isn't enough genuine talent for us to take notice," to George Orwell for *Animal Farm* because "it is impossible to sell animal stories in the U.S.," and to Isaac Bashevis Singer because "it's Poland and the rich Jews again." Before he hit it big, Tony Hillerman's agent dumped him, advising that he should "get rid of all that Indian stuff."[7]

Those were not isolated misjudgments. In fact, many books destined for great success had to survive not just rejection, but repeated rejection. For example, few books today are considered to have more obvious and nearly universal appeal than the works of John Grisham, Theodor Geisel (Dr. Seuss), and J. K. Rowling. Yet the manuscripts they wrote before they became famous—all eventually hugely successful—were all repeatedly rejected. John Grisham's manuscript for *A Time to Kill* was rejected by twenty-six publishers; his second manuscript, for *The Firm*, drew interest from publishers only after a bootleg copy circulating in Hollywood drew a $600,000 offer for the movie rights. Dr. Seuss's first children's book, *And to Think That I Saw It on Mulberry Street*, was rejected by twenty-seven publishers. And J. K. Rowling's first *Harry Potter* manuscript was rejected by nine.[8] Then there is the other side of the coin—the side anyone in the business knows all too well: the many authors who had great potential but never made it, John Grishams who quit after the first twenty rejections or J. K. Rowlings who gave up after the first five. After his many rejections, one such writer, John Kennedy Toole, lost hope of ever getting his novel published and committed suicide. His mother persevered, however, and eleven years later *A Confederacy of Dunces* was published; it won the Pulitzer Prize for Fiction and has sold nearly 2 million copies.

There exists a vast gulf of randomness and uncertainty between the creation of a great novel—or piece of jewelry or chocolate-chip cookie—and the presence of huge stacks of that novel—or jewelry or bags of cookies—at the front of thousands of retail outlets. That's why successful people in every field are almost universally members of a certain set—the set of people who don't give up.

A lot of what happens to us—success in our careers, in our investments, and in our life decisions, both major and minor—is as much the result of random factors as the result of skill, preparedness, and hard work. So the reality that we perceive is not a direct reflection of the people or circumstances that underlie it but is instead an image blurred by the randomizing effects of unforeseeable or fluctuating external forces. That is not to say that ability doesn't matter—it is one of the factors that increase the chances of success—but the connection between actions and results is not as direct as we might like to believe. Thus our past is not so easy to understand, nor is our future so easy to predict, and in both enterprises we benefit from looking beyond the superficial explanations.

WE HABITUALLY UNDERESTIMATE THE EFFECTS of randomness. Our stockbroker recommends that we invest in the Latin American mutual fund that "beat the pants off the domestic funds" five years running. Our doctor attributes that increase in our triglycerides to our new habit of enjoying a Hostess Ding Dong with milk every morning after dutifully feeding the kids a breakfast of mangoes and nonfat yogurt. We may or may not take our stockbroker's or doctor's advice, but few of us question whether he or she has enough data to give it. In the political world, the economic world, the business world—even when careers and millions of dollars are at stake— chance events are often conspicuously misinterpreted as accomplishments or failures.

Hollywood provides a nice illustration. Are the rewards (and punishments) of the Hollywood game deserved, or does luck play a far more important role in box office success (and failure) than people imagine? We all understand that genius doesn't guarantee success, but it's seductive to assume that success must come from genius. Yet the idea that no one can know in advance whether a film will hit or miss has been an uncomfortable suspicion in Hollywood at least since the novelist and screenwriter William Goldman enunciated it in his classic 1983 book *Adventures in the Screen Trade*. In that book, Goldman quoted the former studio executive David Picker as saying, "If I had said yes to all the projects I turned down, and no to all the other ones I took, it would have worked out about the same."[9]

That's not to say that a jittery homemade horror video could become a hit just as easily as, say, *Exorcist: The Beginning*, which cost an estimated $80 million. Well, actually, that is what happened some years back with *The Blair Witch Project:* it cost the filmmakers a mere $60,000 but brought in $140 million in domestic box office revenue—more than three times the business of *Exorcist*. Still, that's not what Goldman was saying. He was referring only to professionally made Hollywood films with production values good enough to land the film a respectable distributor. And Goldman didn't deny that there are reasons for a film's box office performance. But he did say that those reasons are so complex and the path from green light to opening weekend so vulnerable to unforeseeable and uncontrollable influences that educated guesses about an unmade film's potential aren't much better than flips of a coin.

Examples of Hollywood's unpredictability are easy to find. Movie buffs will remember the great expectations the studios had for the megaflops *Ishtar* (Warren Beatty + Dustin Hoffman + a $55 million budget = $14 million in box office revenue) and *Last Action Hero* (Arnold Schwarzenegger + $85 million = $50 million). On the other hand, you might recall the grave doubts that executives at Universal Studios had about the young director George Lucas's film *American Graffiti*, shot for less than $1 million. Despite their skepticism, it took in $115 million, but still that didn't stop them from having even graver doubts about Lucas's next idea. He called the story *Adventures of Luke Starkiller as taken from "The Journal of the Whills."* Universal called it unproducible. Ultimately 20th Century Fox made the film, but the studio's faith in the project went only so far: it paid Lucas just $200,000 to write and direct it; in exchange, Lucas received the sequel and merchandising rights. In the end, *Star Wars* took in

$461 million on a budget of $13 million, and Lucas had himself an empire.

Given the fact that green light decisions are made years before a film is completed and films are subject to many unpredictable factors that arise during those years of production and marketing, not to mention the inscrutable tastes of the audience, Goldman's theory doesn't seem at all far-fetched. (It is also one that is supported by much recent economic research.)[10] Despite all this, studio executives are not judged by the bread-and-butter management skills that are as essential to the head of the United States Steel Corporation as they are to the head of Paramount Pictures. Instead, they are judged by their ability to pick hits. If Goldman is right, that ability is mere illusion, and in spite of his or her swagger no executive is worth that $25 million contract.

Deciding just how much of an outcome is due to skill and how much to luck is not a no-brainer. Random events often come like the raisins in a box of cereal—in groups, streaks, and clusters. And although Fortune is fair in potentialities, she is not fair in outcomes. That means that if each of 10 Hollywood executives tosses 10 coins, although each has an equal chance of being the winner or the loser, in the end there *will* be winners and losers. In this example, the chances are 2 out of 3 that at least 1 of the executives will score 8 or more heads or tails.

Imagine that George Lucas makes a new *Star Wars* film and in one test market decides to perform a crazy experiment. He releases the identical film under two titles: *Star Wars: Episode* A and *Star Wars: Episode B*. Each film has its own marketing campaign and distribution schedule, with the corresponding details identical except that the trailers and ads for one film say *Episode* A and those for the other, *Episode B*. Now we make a contest out of it. Which film will be more popular? Say we look at the first 20,000 moviegoers and record the film they choose to see (ignoring those die-hard fans who will go to both and then insist there were subtle but meaningful differences between the two). Since the films and their marketing campaigns are identical, we can mathematically model the game this way: Imagine lining up all the viewers in a row and flipping a coin for each viewer in turn. If the coin lands heads up, he or she sees *Episode A*; if the coin lands tails up, it's *Episode B*. Because the coin has an equal chance of coming up either way, you might think that in this experimental box office war each film should be in the lead about half the time. But the mathematics of randomness says otherwise: the most probable number of changes in the lead is 0, and it is 88 times more probable that one of the two films will lead through all 20,000 customers than it is that, say, the lead continuously seesaws.[11] The lesson is not that there is no difference between films but that some films will do better than others even if all the films are identical.

Such issues are not discussed in corporate boardrooms, in Hollywood, or elsewhere, and so the typical patterns of randomness—apparent hot or cold streaks or the bunching of data into clusters—are routinely misinterpreted and, worse, acted on as if they represented a new trend.

One of the most high profile examples of anointment and regicide in modern Hollywood was the case of Sherry Lansing, who ran Paramount with great success for many years.[12] Under Lansing, Paramount won Best Picture awards for *Forrest Gump, Braveheart*, and *Titanic* and posted its two highest-grossing years ever. Then Lansing's reputation suddenly plunged, and she was dumped after Paramount experienced, as *Variety* put it, "a long stretch of underperformance at the box office."[13]

In mathematical terms there is both a short and a long explanation for Lansing's fate. First, the short answer. Look at this series of percentages: 11.4, 10.6, 11.3, 7.4, 7.1, 6.7. Notice something? Lansing's boss, Sumner Redstone, did too, and for him the trend was significant, for those six numbers represented the market share of Paramount's Motion Picture Group for the final six years of Lansing's tenure. The trend caused *BusinessWeek* to speculate that Lansing "may simply no longer have Hollywood's hot hand."[14] Soon Lansing announced she was leaving, and a few months later a talent manager named Brad Grey was brought on board.

How can a surefire genius lead a company through seven great years and then fail practically overnight? There were plenty of theories explaining Lansing's early success. While Paramount was doing well, Lansing was praised for making it one of Hollywood's best-run studios and for her knack for turning conventional stories into $100 million hits. When her fortune changed, the revisionists took over. Her penchant for making successful remakes and sequels became a drawback. Most damning of all, perhaps, was the notion that her failure was due to her "middle-of-the-road tastes." She was now blamed for green-lighting such box office dogs as *Timeline* and *Lara Croft Tomb Raider: The Cradle of Life.* Suddenly the conventional wisdom was that Lansing was risk averse, old-fashioned, and out of touch with the trends. But can she really be blamed for thinking that a Michael Crichton bestseller would be promising movie fodder? And where were all the *Lara Croft* critics when the first *Tomb Raider* film took in $131 million in box office revenue?

Even if the theories of Lansing's shortcomings were plausible, consider how abruptly her demise occurred. Did she become risk averse and out of touch overnight? Because Paramount's market share plunged that suddenly. One year Lansing was flying high; the next she was a punch line for late-night comedians. Her change of fortune might have been understandable if, like others in Hollywood, she had become depressed over a nasty divorce proceeding, had been charged with embezzlement, or had joined a religious cult. That was not the case. And she certainly hadn't sustained any damage to her cerebral cortex. The only evidence of Lansing's newly developed failings that her critics could offer was, in fact, her newly developed failings.

In hindsight it is clear that Lansing was fired because of the industry's misunderstanding of randomness and not because of her flawed decision making: Paramount's films for the following year were already in the pipeline when Lansing left the company. So if we want to know roughly how Lansing would have done in some parallel universe in which she remained in her job, all we need to do is look at the data in the year following her departure. With such films as *War of the Worlds* and *The Longest Yard*, Paramount had its best summer in a decade and saw its market share rebound to nearly 10 percent. That isn't merely ironic — it's again that aspect of randomness called regression toward the mean. A *Variety* headline on the subject read, "Parting Gifts: Old Regime's Pics Fuel Paramount Rebound,"[15] but one can't help but think that had Viacom (Paramount's parent company) had more patience, the headline might have read, "Banner Year Puts Paramount and Lansing's Career Back on Track."

Sherry Lansing had good luck at the beginning and bad luck at the end, but it could have been worse. She could have had her bad luck at the beginning. That's what happened to a Columbia Pictures chief named Mark Canton. Described as box office savvy and enthusiastic shortly after he was hired, he was fired after his first few years produced disappointing box office results. Criticized by one unnamed colleague for being "incapable of distinguishing the winners from the losers" and by another for being "too busy cheerleading," this disgraced man left in the pipeline when he departed such films as *Men in Black* ($589 million in worldwide box office revenue), *Air Force One* ($315 million), *The Fifth Element* ($264 million), *Jerry Maguire* ($274 million), and *Anaconda* ($137 million). As *Variety* put it, Canton's legacy pictures "hit and hit big."[16]

Well, that's Hollywood, a town where Michael Ovitz works as Disney president for fifteen months and then leaves with a $140 million severance package and where the studio head David Begelman is fired by Columbia Pictures for forgery and embezzlement and then is hired a few years later as CEO of MGM. But as we'll see in the following chapters, the same sort of misjudgments that plague Hollywood also plague people's perceptions in all realms of life.

MY OWN EPIPHANY regarding the hidden effects of randomness came in college, when I took a course in probability and began applying its principles to the sports world. That is easy to do because, as in the film business, most accomplishments in sports are easily quantified and the data are readily available. What I discovered was that just as the lessons of persistence, practice, and teamwork that we learn from sports apply equally to all endeavors of life, so do the lessons of randomness. And so I set out to examine a tale of two baseball sluggers, Roger Maris and Mickey Mantle, a tale that bears a lesson for all of us, even those who wouldn't know a baseball from a Ping-Pong ball.

The year was 1961. I was barely of reading age, but I still recall the faces of Maris and his more popular New York Yankees teammate, Mantle, on the cover of *Life* magazine. The two baseball players were engaged in a historic race to tie or break Babe Ruth's beloved 1927 record of 60 home runs in one year. Those were idealistic times when my teacher would say things like "we need more heroes like Babe Ruth," or "we never had a crooked president." Because the legend of Babe Ruth was sacred, anyone who might challenge it had better be worthy. Mantle, a courageous perennial slugger who fought on despite bad knees, was the fans'—and the press's—overwhelming favorite. A good-looking, good-natured fellow, Mantle came across as the kind of all-American boy everyone hoped would set records. Maris, on the other hand, was a gruff, private fellow, an underdog who had never hit more than 39 home runs in a year, much less anywhere near 60. He was seen as a nasty sort, someone who didn't give interviews and didn't like kids. They all rooted for Mantle. I liked Maris.

As it turned out, Mantle's knees got the best of him, and he made it to only 54 home runs. Maris broke Ruth's record with 61. Over his career, Babe Ruth had hit 50 or more home runs in a season four times and twelve times had hit more than anyone else in the league. Maris never again hit 50 or even 40 and never again led the league. That overall performance fed the resentment. As the years went by, Maris was criticized relentlessly by fans, sportswriters, and sometimes other players. Their verdict: he had crumbled under the pressure of being a champion. Said one famous baseball old-timer, "Maris had no right to break Ruth's record."[17] That may have been true, but not for the reason the old-timer thought.

Many years later, influenced by that college math course, I would

learn to think about Maris's achievement in a new light. To analyze
the Ruth-Mantle race I reread that old *Life* article and found in it a
brief discussion of probability theory[18] and how it could be used to
predict the result of the Maris-Mantle race. I decided to make my
own mathematical model of home run hitting. Here's how it goes:
The result of any particular at bat (that is, an opportunity for success)
depends primarily on the player's ability, of course. But it also
depends on the interplay of many other factors: his health; the wind,
the sun, or the stadium lights; the quality of the pitches he receives;
the game situation; whether he correctly guesses how the pitcher will
throw; whether his hand-eye coordination works just perfectly as he
takes his swing; whether that brunette he met at the bar kept him up
too late or the chili-cheese dog with garlic fries he had for breakfast
soured his stomach. If not for all the unpredictable factors, a player
would either hit a home run on every at bat or fail to do so. Instead,
for each at bat all you can say is that he has a certain probability of
hitting a home run and a certain probability of failing to hit one.
Over the hundreds of at bats he has each year, those random factors
usually average out and result in some typical home run production
that increases as the player becomes more skillful and then eventually decreases owing to the same process that etches wrinkles in his
handsome face. But sometimes the random factors don't average out.
How often does that happen, and how large is the aberration?

From the player's yearly statistics you can estimate his probability
of hitting a home run at each opportunity—that is, on each trip to the
plate.[19] In 1960, the year before his record year, Roger Maris hit
1 home run for every 14.7 opportunities (about the same as his home
run output averaged over his four prime years). Let's call this performance normal Maris. You can model the home run hitting skill of
normal Maris this way: Imagine a coin that comes up heads on average not 1 time every 2 tosses but 1 time every 14.7. Then flip that
coin 1 time for every trip to the plate and award Maris 1 home run
every time the coin comes up heads. If you want to match, say,
Maris's 1961 season, you flip the coin once for every home run
opportunity he had that year. By that method you can generate a

whole series of alternative 1961 seasons in which Maris's skill level matches the home run totals of normal-Maris. The results of those mock seasons illustrate the range of accomplishment that normal Maris could have expected in 1961 if his talent had not spiked—that is, given only his "normal" home run ability plus the effects of pure luck.

To have actually performed this experiment, I'd have needed a rather odd coin, a rather strong wrist, and a leave of absence from college. In practice the mathematics of randomness allowed me to do the analysis employing equations and a computer. In most of my imaginary 1961 seasons, normal Maris's home run output was, not surprisingly, in the range that was normal for Maris. Some mock seasons he hit a few more, some a few less. Only rarely did he hit a lot more or a lot less. How frequently did normal Maris's talent produce Ruthian results?

I had expected normal Maris's chances of matching Ruth's record to be roughly equal to Jack Whittaker's when he plopped down an extra dollar as he bought breakfast biscuits at a convenience store a few years back and ended up winning $314 million in his state Powerball lottery. That's what a less talented player's chances would have been. But normal Maris, though not Ruthian, was still far above average at hitting home runs. And so normal Maris's probability of producing a record output by chance was not microscopic: he matched or broke Ruth's record about 1 time every 32 seasons. That might not sound like good odds, and you probably wouldn't have wanted to bet on either Maris or the year 1961 in particular. But those odds lead to a striking conclusion. To see why, let's now ask a more interesting question. Let's consider *all* players with the talent of normal Maris and the *entire* seventy-year period from Ruth's record to the start of the "steroid era" (when, because of players' drug use, home runs became far more common). What are the odds that *some* player at *some* time would have matched or broken Ruth's record by chance alone? Is it reasonable to believe that Maris just happened to be the recipient of the lucky aberrant season?

History shows that in that period there was about 1 player every

3 years with both the talent and the opportunities comparable to those of normal Maris in 1961. When you add it all up, that makes the probability that by chance alone one of those players would have matched or broken Ruth's record a little greater than 50 percent. In other words, over a period of seventy years a random spike of 60 or more home runs for a player whose production process merits more like 40 home runs is to be expected—a phenomenon something like that occasional loud crackle you hear amid the static in a bad telephone connection. It is also to be expected, of course, that we will deify, or vilify—and certainly endlessly analyze—whoever that "lucky" person turns out to be.

We can never know for certain whether Maris was a far better player in 1961 than in any of the other years he played professional baseball or whether he was merely the beneficiary of good fortune. But detailed analyses of baseball and other sports by scientists as eminent as the late Stephen Jay Gould and the Nobel laureate E. M. Purcell show that coin-tossing models like the one I've described match very closely the actual performance of both players and teams, including their hot and cold streaks.[20]

When we look at extraordinary accomplishments in sports—or elsewhere—we should keep in mind that extraordinary events can happen without extraordinary causes. Random events often look like nonrandom events, and in interpreting human affairs we must take care not to confuse the two. Though it has taken many centuries, scientists have learned to look beyond apparent order and recognize the hidden randomness in both nature and everyday life. In this chapter I've presented a few glimpses of those workings. In the following chapters I shall consider the central ideas of randomness within their historical context and describe their relevance with the aim of offering a new perspective on our everyday surroundings and hence a better understanding of the connection between this fundamental aspect of nature and our own experience.

CHAPTER 2

The Laws of Truths
and Half-Truths

L OOKING TO THE SKY on a clear, moonless night, the human eye can detect thousands of twinkling sources of light. Nestled among those haphazardly scattered stars are patterns. A lion here, a dipper there. The ability to detect patterns can be both a strength and a weakness. Isaac Newton pondered the patterns of falling objects and created a law of universal gravitation. Others have noted a spike in their athletic performance when they are wearing dirty socks and thenceforth have refused to wear clean ones. Among all the patterns of nature, how do we distinguish the meaningful ones? Drawing that distinction is an inherently practical enterprise. And so it might not astonish you to learn that, unlike geometry, which arose as a set of axioms, proofs, and theorems created by a culture of ponderous philosophers, the theory of randomness sprang from minds focused on spells and gambling, figures we might sooner imagine with dice or a potion in hand than a book or a scroll.

The theory of randomness is fundamentally a codification of common sense. But it is also a field of subtlety, a field in which great experts have been famously wrong and expert gamblers infamously correct. What it takes to understand randomness and overcome our misconceptions is both experience and a lot of careful thinking. And so we begin our tour with some of the basic laws of probability and the challenges involved in uncovering, understanding, and applying them. One of the classic explorations of people's intuition about those laws was an experiment conducted by the pair who did so much to elucidate our misconceptions, Daniel Kahneman and Amos Tversky.[1] Feel free to take part—and learn something about your own probabilistic intuition.

Imagine a woman named Linda, thirty-one years old, single, outspoken, and very bright. In college she majored in philosophy. While a student she was deeply concerned with discrimination and social justice and participated in antinuclear demonstrations. Tversky and Kahneman presented this description to a group of eighty-eight subjects and asked them to rank the following statements on a scale of 1 to 8 according to their probability, with 1 representing the most probable and 8 the least. Here are the results, in order from most to least probable:

Statement	Average Probability Rank
Linda is active in the feminist movement.	2.1
Linda is a psychiatric social worker.	3.1
Linda works in a bookstore and takes yoga classes.	3.3
Linda is a bank teller and is active in the feminist movement.	4.1
Linda is a teacher in an elementary school.	5.2
Linda is a member of the League of Women Voters.	5.4
Linda is a bank teller.	6.2
Linda is an insurance salesperson.	6.4

At first glance there may appear to be nothing unusual in these results: the description was in fact designed to be representative of an active feminist and unrepresentative of a bank teller or an insurance salesperson. But now let's focus on just three of the possibilities and their average ranks, listed below in order from most to least probable. This is the order in which 85 percent of the respondents ranked the three possibilities:

Statement	Average Probability Rank
Linda is active in the feminist movement.	2.1
Linda is a bank teller and is active in the feminist movement.	4.1
Linda is a bank teller.	6.2

If nothing about this looks strange, then Kahneman and Tversky have fooled you, for if the chance that Linda is a bank teller and is active in the feminist movement were greater than the chance that Linda is a bank teller, there would be a violation of our first law of probability, which is one of the most basic of all: *The probability that two events will both occur can never be greater than the probability that each will occur individually.* Why not? Simple arithmetic: the chances that event A will occur = the chances that events A and B will occur + the chance that event A will occur and event B *will not* occur.

Kahneman and Tversky were not surprised by the result because they had given their subjects a large number of possibilities, and the connections among the three scenarios could easily have gotten lost in the shuffle. And so they presented the description of Linda to another group, but this time they presented only these possibilities:

Linda is active in the feminist movement.
Linda is a bank teller and is active in the feminist movement.
Linda is a bank teller.

To their surprise, 87 percent of the subjects in this trial also ranked the probability that Linda is a bank teller and is active in the feminist movement higher than the probability that Linda is a bank teller. And so the researchers pushed further: they explicitly asked a group of thirty-six fairly sophisticated graduate students to consider their answers in light of our first law of probability. Even after the prompting, two of the subjects clung to the illogical response.

The interesting thing that Kahneman and Tversky noticed about this stubborn misperception is that people will not make the same mistake if you ask questions that are unrelated to what they know about Linda. For example, suppose Kahneman and Tversky had asked which of these statements seems most probable:

Linda owns an International House of Pancakes franchise.
Linda had a sex-change operation and is now known as Larry.
Linda had a sex-change operation, is now known as Larry,
    and owns an International House of Pancakes franchise.

In this case few people would choose the last option as more likely than either of the other two.

Kahneman and Tversky concluded that because the detail "Linda is active in the feminist movement" rang true based on the initial description of her character, when they added that detail to the bank-teller speculation, it increased the scenario's credibility. But a lot could have happened between Linda's hippie days and her fourth decade on the planet. She might have undergone a conversion to a fundamentalist religious cult, married a skinhead and had a swastika tattooed on her left buttock, or become too busy with other aspects of her life to remain politically active. In each of these cases and many others she would probably not be active in the feminist movement. So adding that detail lowered the chances that the scenario was accurate even though it appeared to raise the chances of its accuracy.

If the details we are given fit our mental picture of something, then the more details in a scenario, the more real it seems and hence the more probable we consider it to be—even though any act of adding less-than-certain details to a conjecture makes the conjecture less probable. This inconsistency between the logic of probability and people's assessments of uncertain events interested Kahneman and Tversky because it can lead to unfair or mistaken assessments in real-life situations. Which is more likely: that a defendant, after discovering the body, left the scene of the crime or that a defendant, after discovering the body, left the scene of the crime because he feared being accused of the grisly murder? Is it more probable that the president will increase federal aid to education or that he or she will increase federal aid to education with funding freed by cutting other aid to the states? Is it more likely that your company will increase sales next year or that it will increase sales next year because the overall economy has had a banner year? In each case, even though the latter is less probable than the former, it may sound more likely. Or as Kahneman and Tversky put it, "A good story is often less probable than a less satisfactory . . . [explanation]."

Kahneman and Tversky found that even highly trained doctors make this error.[2] They presented a group of internists with a serious medical problem: a pulmonary embolism (a blood clot in the lung). If you have that ailment, you might display one or more of a set of symptoms. Some of those symptoms, such as partial paralysis, are uncommon; others, such as shortness of breath, are probable. Which is more likely: that the victim of an embolism will experience only partial paralysis or that the victim will experience both partial paralysis and shortness of breath? Kahneman and Tversky found that 91 percent of the doctors believed a clot was less likely to cause just a rare symptom than it was to cause a combination of the rare symptom and a common one. (In the doctors' defense, patients don't walk into their offices and say things like "I have a blood clot in my lungs. Guess my symptoms.")

Years later one of Kahneman's students and another researcher found that attorneys fall prey to the same bias in their judgments.[3] Whether involved in a criminal case or a civil case, clients typically depend on their lawyers to assess what may occur if their case goes to trial. What are the chances of acquittal or of a settlement or a monetary judgment in various amounts? Although attorneys might not phrase their opinions in terms of numerical probabilities, they offer advice based on their personal forecast of the relative likelihood of the possible outcomes. Here, too, the researchers found that lawyers assign higher probabilities to contingencies that are described in greater detail. For example, at the time of the civil lawsuit brought by Paula Jones against then president Bill Clinton, 200 practicing lawyers were asked to predict the probability that the trial would not run its full course. For some of the subjects that possibility was broken down into specific causes for the trial's early end, such as settlement, withdrawal of the charges, or dismissal by the judge. In comparing the two groups—lawyers who had simply been asked to predict whether the trial would run its full course and lawyers who had been presented with ways in which the trial might reach a premature conclusion—the researchers found that the lawyers who had been presented with causes of a premature conclusion were much more likely than the other lawyers to predict that the trial would reach an early end.

The ability to evaluate meaningful connections among different phenomena in our environment may be so important that it is worth seeing a few mirages. If a starving caveman sees an indistinct greenish blur on a distant rock, it is more costly to dismiss it as uninteresting when it is in reality a plump, tasty lizard than it is to race over and pounce on what turns out to be a few stray leaves. And so, that theory goes, we might have evolved to avoid the former mistake at the cost of sometimes making the latter.

IN THE STORY of mathematics the ancient Greeks stand out as the inventors of the manner in which modern mathematics is carried out: through axioms, proofs, theorems, more proofs, more theorems, and so on. In the 1930s, however, the Czech American mathematician Kurt Gödel—a friend of Einstein's—showed this approach to be somewhat deficient: most of mathematics, he demonstrated, must be inconsistent or else must contain truths that cannot be proved. Still, the march of mathematics has continued unabated in the Greek style, the style of Euclid. The Greeks, geniuses in geometry, created a small set of axioms, statements to be accepted without proof, and proceeded from there to prove many beautiful theorems detailing the properties of lines, planes, triangles, and other geometric forms. From this knowledge they discerned, for example, that the earth is a sphere and even calculated its radius. One must wonder why a civilization that could produce a theorem such as proposition 29 of book 1 of Euclid's *Elements*—"a straight line falling on two parallel straight lines makes the alternate angles equal to one another, the exterior angle equal to the interior and opposite angle, and the interior angles on the same side equal to two right angles"—did not create a theory showing that if you throw two dice, it would be unwise to bet your Corvette on their both coming up a 6.

Actually, not only didn't the Greeks have Corvettes, but they also didn't have dice. They did have gambling addictions, however. They also had plenty of animal carcasses, and so what they tossed were astragali, made from heel bones. An astragalus has six sides, but only four are stable enough to allow the bone to come to rest on them. Modern scholars note that because of the bone's construction, the chances of its landing on each of the four sides are not equal: they are about 10 percent for two of the sides and 40 percent for the other two. A common game involved tossing four astragali. The outcome considered best was a rare one, but not the rarest: the case in which all four astragali came up different. This was called a Venus throw. The Venus throw has a probability of about 384 out of 10,000, but the Greeks, lacking a theory of randomness, didn't know that.

The Greeks also employed astragali when making inquiries of their oracles. From their oracles, questioners could receive answers that were said to be the words of the gods. Many important choices made by prominent Greeks were based on the advice of oracles, as evidenced by the accounts of the historian Herodotus, and writers like Homer, Aeschylus, and Sophocles. But despite the importance of astragali tosses in both gambling and religion, the Greeks made no effort to understand the regularities of astragali throws.

Why didn't the Greeks develop a theory of probability? One answer is that many Greeks believed that the future unfolded according to the will of the gods. If the result of an astragalus toss meant "marry the stocky Spartan girl who pinned you in that wrestling match behind the school barracks," a Greek boy wouldn't view the toss as the lucky (or unlucky) result of a random process; he would view it as the gods' will. Given such a view, an understanding of randomness would have been irrelevant. Thus a mathematical prediction of randomness would have seemed impossible. Another answer may lie in the very philosophy that made the Greeks such great mathematicians: they insisted on absolute truth, proved by logic and axioms, and frowned on uncertain pronouncements. In Plato's *Phaedo,* for example, Simmias tells Socrates that "arguments from probabilities are impostors" and anticipates the work of Kahneman and Tversky by pointing out that "unless great caution is observed in the use of them they are apt to be deceptive—in geometry, and in other things too."[4] And in *Theaetetus,* Socrates says that any mathematician "who argued from probabilities and likelihoods in geometry would not be worth an ace."[5] But even Greeks who believed that probabilists were worth an ace might have had difficulty working out a consistent theory in those days before extensive record keeping because people have notoriously poor memories when it comes to estimating the frequency—and hence the probability—of past occurrences.

Which is greater: the number of six-letter English words having *n* as their fifth letter or the number of six-letter English words ending in *ing?* Most people choose the group of words ending in *ing.*[6] Why? Because words ending in *ing* are easier to think of than generic six-letter words having *n* as their fifth letter. But you don't have to survey the *Oxford English Dictionary*—or even know how to count—to prove that guess wrong: the group of six-letter words having *n* as their fifth letter words *includes* all six-letter words ending in *ing.* Psychologists call that type of mistake the availability bias because in reconstructing the past, we give unwarranted importance to memories that are most vivid and hence most available for retrieval.

The nasty thing about the availability bias is that it insidiously distorts our view of the world by distorting our perception of past events and our environment. For example, people tend to overestimate the fraction of homeless people who are mentally ill because when they encounter a homeless person who is not behaving oddly, they don't take notice and tell all their friends about that unremarkable homeless person they ran into. But when they encounter a homeless person stomping down the street and waving his arms at an imaginary companion while singing "When the Saints Go Marching In," they do tend to remember the incident.[7] How probable is it that of the five lines at the grocery-store checkout you will choose the one that takes the longest? Unless you've been cursed by a practitioner of the black arts, the answer is around 1 in 5. So why, when you look back, do you get the feeling you have a supernatural knack for choosing the longest line? Because you have more important things to focus on when things go right, but it makes an impression when the lady in front of you with a single item in her cart decides to argue about why her chicken is priced at $1.50 a pound when she is certain the sign at the meat counter said $1.49.

One stark illustration of the effect the availability bias can have on our judgment and decision making came from a mock jury trial.[8] In the study the jury was given equal doses of exonerating and incriminating evidence regarding the charge that a driver was drunk when he ran into a garbage truck. The catch is that one group of jurors was given the exonerating evidence in a "pallid" version: "The owner of the garbage truck stated under cross-examination that his garbage truck was difficult to see at night because it was gray in color." The other group was given a more "vivid" form of the same evidence: "The owner of the garbage truck stated under cross-examination that his garbage truck was difficult to see at night because it was gray in color. The owner remarked his trucks are gray 'because it hides the dirt. What do you want, I should paint 'em pink?' " The incriminating evidence was also presented in two ways, this time in a vivid form to the first group and in a pallid version to the second. When the jurors were asked to produce guilt/innocence ratings, the side with the more vivid presentation of the evidence always prevailed, and the effect was enhanced when there was a forty-eight-hour delay before rendering the verdict (presumably because the recall gap was even greater).

By distorting our view of the past, the availability bias complicates any attempt to make sense of it. That was true for the ancient Greeks just as it is true for us. But there was one other major obstacle to an

early theory of randomness, a very practical one: although basic probability requires only knowledge of arithmetic, the Greeks did not
know arithmetic, at least not in a form that is easy to work with. In
Athens in the fifth century B.C., for instance, at the height of Greek
civilization, a person who wanted to write down a number used a
kind of alphabetic code.[9] The first nine of the twenty-four letters in
the Greek alphabet stood for the numbers we call 1 through 9. The
next nine letters stood for the numbers we call 10, 20, 30, and so on.
And the last six letters plus three additional symbols stood for the first
nine hundreds (100, 200, and so on, to 900). If you think you have
trouble with arithmetic now, imagine trying to subtract $\Delta\Gamma\Theta$ from
$\Omega\Psi\Pi$! To make matters worse, the order in which the ones, tens, and
hundreds were written didn't really matter: sometimes the hundreds
were written first, sometimes last, and sometimes all order was
ignored. Finally, the Greeks had no zero.

The concept of zero came to Greece when Alexander invaded the
Babylonian Empire in 331 B.C. Even then, although the Alexandri-
ans began to use the zero to denote the absence of a number, it
wasn't employed as a number in its own right. In modern mathematics the number 0 has two key properties: in addition it is the number
that, when added to any other number, leaves the other number
unchanged, and in multiplication it is the number that, when multiplied by any other number, is itself unchanged. This concept wasn't
introduced until the ninth century, by the Indian mathematician
Mahāvīra.

Even after the development of a usable number system it would
be many more centuries before people came to recognize addition,
subtraction, multiplication, and division as the fundamental arithmetic operations—and slowly realized that convenient symbols
would make their manipulation far easier. And so it wasn't until the
sixteenth century that the Western world was truly poised to develop
a theory of probability. Still, despite the handicap of an awkward system of calculation, it was the civilization that conquered the
Greeks—the Romans—who made the first progress in understanding
randomness.

THE ROMANS generally scorned mathematics, at least the mathematics of the Greeks. In the words of the Roman statesman Cicero, who lived from 106 to 43 B.C., "The Greeks held the geometer in the highest honor; accordingly, nothing made more brilliant progress among them than mathematics. But we have established as the limit of this art its usefulness in measuring and counting."[10] Indeed, whereas one might imagine a Greek textbook focused on the proof of congruences among abstract triangles, a typical Roman text focused on such issues as how to determine the width of a river when the enemy is occupying the other bank.[11] With such mathematical priorities, it is not surprising that while the Greeks produced mathematical luminaries like Archimedes, Diophantus, Euclid, Eudoxus, Pythagoras, and Thales; the Romans did not produce even one mathematician.[12] In Roman culture it was comfort and war, not truth and beauty, that occupied center stage. And yet precisely because they focused on the practical, the Romans saw value in understanding probability. So while finding little value in abstract geometry, Cicero wrote that "probability is the very guide of life."[13]

Cicero was perhaps the greatest ancient champion of probability. He employed it to argue against the common interpretation of gambling success as due to divine intervention, writing that the "man who plays often will at some time or other make a Venus cast: now and then indeed he will make it twice and even thrice in succession. Are we going to be so feeble-minded then as to affirm that such a thing happened by the personal intervention of Venus rather than by pure luck?"[14] Cicero believed that an event could be anticipated and predicted even though its occurrence would be a result of blind chance. He even used a statistical argument to ridicule the belief in astrology. Annoyed that although outlawed in Rome, astrology was nevertheless alive and well, Cicero noted that at Cannae in 216 B.C., Hannibal, leading about 50,000 Carthaginian and allied troops, crushed the much larger Roman army, slaughtering more than 60,000 of its 80,000 soldiers. "Did all the Romans who fell at Cannae have the same horoscope?" Cicero asked. "Yet all had one and the same end."[15] Cicero might have been encouraged to know that a couple of thousand years later in the journal *Nature* a scientific study of the validity of astrological predictions agreed with his conclusion.[16] The *New York Post*, on the other hand, advises today that as a Sagittarius, I must look at criticisms objectively and make whatever changes seem necessary.

In the end, Cicero's principal legacy in the field of randomness is the term he used, *probabilis*, which is the origin of the term we employ today. But it is one part of the Roman code of law, the Digest, compiled by Emperor Justinian in the sixth century, that is the first document in which probability appears as an everyday term of art.[17] To appreciate the Roman applications of mathematical thinking to legal theory, one must understand the context: Roman law in the Dark Ages was based on the practice of the Germanic tribes. It wasn't pretty. Take, for example, the rules of testimony. The veracity of, say, a husband denying an affair with his wife's toga maker would be determined not by hubby's ability to withstand a grilling by prickly opposing counsel but by whether he'd stick to his story even after being pricked—by a red-hot iron. (Bring back *that* custom and you'll see a lot more divorce cases settled out of court.) And if the defendant says the chariot never tried to stop but the expert witness says the hoof prints show that the brakes were applied, Germanic doctrine offered a simple prescription: "Let one man be chosen from each group to fight it out with shields and spears. Whoever loses is a perjurer and must lose his right hand."[18]

In replacing, or at least supplementing, the practice of trial by battle, the Romans sought in mathematical precision a cure for the deficiencies of their old, arbitrary system. Seen in this context, the Roman idea of justice employed advanced intellectual concepts. Recognizing that evidence and testimony often conflicted and that the best way to resolve such conflicts was to quantify the inevitable uncertainty, the Romans created the concept of half proof, which applied in cases in which there was no compelling reason to believe or disbelieve evidence or testimony. In some cases the Roman doctrine of evidence included even finer degrees of proof, as in the church decree that "a bishop should not be condemned except with seventy-two witnesses . . . a cardinal priest should not be condemned except with forty-four witnesses, a cardinal deacon of the city of Rome without thirty-six witnesses, a subdeacon, acolyte, exorcist, lector, or doorkeeper except with seven witnesses."[19] To be convicted under those rules, you'd have to have not only committed the crime but also sold tickets. Still, the recognition that the probability of truth in testimony can vary and that rules for combining such probabilities are necessary was a start. And so it was in the unlikely venue of ancient Rome that a systematic set of rules based on probability first arose.

Unfortunately it is hard to achieve quantitative dexterity when you're juggling VIIIs and XIVs. In the end, though Roman law had a certain legal rationality and coherence, it fell short of mathematical validity. In Roman law, for example, two half proofs constituted a complete proof. That might sound reasonable to a mind unaccustomed to quantitative thought, but with today's familiarity with fractions it invites the question, if two half proofs equal a complete certainty, what do three half proofs make? According to the correct manner of compounding probabilities, not only do two half proofs yield less than a whole certainty, but no finite number of partial proofs will ever add up to a certainty because to compound probabilities, you don't add them; you multiply.

That brings us to our next law, the rule for compounding probabilities: *If two possible events, A and B, are independent, then the probability that both A and B will occur is equal to the product of their individual probabilities.* Suppose a married person has on average roughly a 1 in 50 chance of getting divorced each year. On the other hand, a police officer has about a 1 in 5,000 chance each year of being killed on the job. What are the chances that a married police officer will be divorced and killed in the same year? According to the above principle, if those events were independent, the chances would be roughly $\frac{1}{50} \times \frac{1}{5,000}$, which equals $\frac{1}{250,000}$. Of course the events are not independent; they are linked: once you die, darn it, you can no longer get divorced. And so the chance of that much bad luck is actually a little less than 1 in 250,000.

Why multiply rather than add? Suppose you make a pack of trading cards out of the pictures of those 100 guys you've met so far through your Internet dating service, those men who in their Web site photos often look like Tom Cruise but in person more often resemble Danny DeVito. Suppose also that on the back of each card you list certain data about the men, such as honest (yes or no) and attractive (yes or no). Finally, suppose that 1 in 10 of the prospective soul mates rates a yes in each case. How many in your pack of 100 will pass the test on both counts? Let's take honest as the first trait (we could equally well have taken attractive). Since 1 in 10 cards lists a yes under honest, 10 of the 100 cards will qualify. Of those 10, how many are attractive? Again, 1 in 10, so now you are left with 1 card. The first 1 in 10 cuts the possibilities down by $\frac{1}{10}$, and so does the next 1 in 10, making the result 1 in 100. That's why you multiply. And if you have more requirements than just honest and attractive, you have to keep multiplying, so . . . well, good luck.

Before we move on, it is worth paying attention to an important detail: the clause that reads *if two possible events, A and B, are independent.* Suppose an airline has 1 seat left on a flight and 2 passengers have yet to show up. Suppose that from experience the airline knows there is a 2 in 3 chance a passenger who books a seat will arrive to claim it. Employing the multiplication rule, the gate attendant can conclude there is a $\frac{2}{3} \times \frac{2}{3}$ or about a 44 percent chance she will have to deal with an unhappy customer. The chance that neither customer will show and the plane will have to fly with an empty seat, on the other hand, is $\frac{1}{3} \times \frac{1}{3}$, or only about 11 percent. But that assumes the passengers are independent. If, say, they are traveling together, then the above analysis is wrong. The chances that both will show up are 2 in 3, the same as the chances that one will show up. It is important to remember that you get the compound probability from the simple ones by multiplying only if the events are in no way contingent on each other.

The rule we just applied could be applied to the Roman rule of half proofs: the chances of two independent half proofs' being wrong are 1 in 4, so two half proofs constitute three-fourths of a proof, not a whole proof. The Romans added where they should have multiplied.

There are situations in which probabilities *should* be added, and that is our next law. It arises when we want to know the chances of either one event or another occurring, as opposed to the earlier situation, in which we wanted to know the chance of one event and another event both happening. The law is this: *If an event can have a number of different and distinct possible outcomes, A, B, C, and so on, then the probability that either A or B will occur is equal to the sum of the individual probabilities of A and B, and the sum of the probabilities of all the possible outcomes (A, B, C, and so on) is 1 (that is, 100 percent).* When you want to know the chances that two independent events, A and B, will both occur, you multiply; if you want to know the chances that either of two mutually exclusive events, A or B, will occur, you add. Back to our airline: when should the gate attendant add the probabilities instead of multiplying them? Suppose she wants to know the chances that either both passengers or neither passenger will show up. In this case she should add the individual probabilities, which according to what we calculated above, would come to 55 percent.

These three laws, simple as they are, form much of the basis of probability theory. Properly applied, they can give us much insight into the workings of nature and the everyday world. We employ them in our everyday decision making all the time. But like the Roman lawmakers, we don't always use them correctly.

IT IS EASY TO LOOK BACK, shake our heads, and write books with titles like *The Rotten Romans* (Scholastic, 1994). But lest we become unjustifiably self-congratulatory, I shall end this chapter with a look at some ways in which the basic laws I've discussed may be applied to our own legal system. As it turns out, that's enough to sober up anyone drunk on feelings of cultural superiority.

The good news is that we don't have half proofs today. But we do have a kind of $^{999,000}/_{1,000,000}$ proof. For instance, it is not uncommon for experts in DNA analysis to testify at a criminal trial that a DNA sample taken from a crime scene matches that taken from a suspect. How certain are such matches? When DNA evidence was first introduced, a number of experts testified that false positives are impossible in DNA testing. Today DNA experts regularly testify that the odds of a random person's matching the crime sample are less than 1 in 1 million or 1 in 1 billion. With those odds one could hardly blame a juror for thinking, *throw away the key*. But there is another statistic that is often not presented to the jury, one having to do with the fact that labs make errors, for instance, in collecting or handling a sample, by accidentally mixing or swapping samples, or by misinterpreting or incorrectly reporting results. Each of these errors is rare but not nearly as rare as a random match. The Philadelphia City Crime Laboratory, for instance, admitted that it had swapped the reference sample of the defendant and the victim in a rape case, and a testing firm called Cellmark Diagnostics admitted a similar error.[20] Unfortunately, the power of statistics relating to DNA presented in court is such that in Oklahoma a court sentenced a man named Timothy Durham to more than 3,100 years in prison even though eleven witnesses had placed him in another state at the time of the crime. It turned out that in the initial analysis the lab had failed to completely separate the DNA of the rapist and that of the victim in the fluid they tested, and the combination of the victim's and the rapist's DNA produced a positive result when compared with Durham's. A later retest turned up the error, and Durham was released after spending nearly four years in prison.[21]

Estimates of the error rate due to human causes vary, but many experts put it at around 1 percent. However, since the error rate of many labs has never been measured, courts often do not allow testimony on this overall statistic. Even if courts did allow testimony regarding false positives, how would jurors assess it? Most jurors assume that given the two types of error—the 1 in 1 billion accidental match and the 1 in 100 lab-error match—the overall error rate must be somewhere in between, say 1 in 500 million, which is still for most jurors beyond a reasonable doubt. But employing the laws of probability, we find a much different answer.

The way to think of it is this: Since both errors are very unlikely, we can ignore the possibility that there is both an accidental match *and* a lab error. Therefore, we seek the probability that one error *or* the other occurred. That is given by our sum rule: it is the probability of a lab error (1 in 100) + the probability of an accidental match (1 in 1 billion). Since the latter is 10 million times smaller than the former, to a very good approximation the chance of both errors is the same as the chance of the more probable error—that is, the chances are 1 in 100. Given both possible causes, therefore, we should ignore the fancy expert testimony about the odds of accidental matches and focus instead on the much higher laboratory error rate—the very data courts often do not allow attorneys to present! And so the oft-repeated claims of DNA infallibility are exaggerated.

This is not an isolated issue. The use of mathematics in the modern legal system suffers from problems no less serious than those that arose in Rome so many centuries ago. One of the most famous cases illustrating the use and misuse of probability in law is *People v. Collins,* heard in 1968 by the California Supreme Court.[22] Here are the facts of the case as presented in the court decision:

> On June 18, 1964, about 11:30 a.m. Mrs. Juanita Brooks, who had been shopping, was walking home along an alley in the San Pedro area of the city of Los Angeles. She was pulling behind her a wicker basket carryall containing groceries and had her purse on top of the packages. She was using a cane. As she stooped down to pick up an empty carton, she was suddenly pushed to the ground by a person whom she neither saw nor heard approach. She was stunned by the fall and felt some pain. She managed to look up and saw a young woman running from the scene. According to Mrs. Brooks the latter appeared to weigh about 145 pounds, was wearing "something dark," and had hair "between a dark blond and a light blond," but lighter than the color of defendant Janet Collins' hair as it

appeared at the trial. Immediately after the incident, Mrs. Brooks discovered that her purse, containing between $35 and $40, was missing.

About the same time as the robbery, John Bass, who lived on the street at the end of the alley, was in front of his house watering his lawn. His attention was attracted by "a lot of crying and screaming" coming from the alley. As he looked in that direction, he saw a woman run out of the alley and enter a yellow automobile parked across the street from him. He was unable to give the make of the car. The car started off immediately and pulled wide around another parked vehicle so that in the narrow street it passed within six feet of Bass. The latter then saw that it was being driven by a male Negro, wearing a mustache and beard. . . . Other witnesses variously described the car as yellow, as yellow with an off-white top, and yellow with an egg-shell white top. The car was also described as being medium to large in size.

A few days after the incident a Los Angeles police officer spotted a yellow Lincoln with an off-white top in front of the defendants' home and spoke with them, explaining that he was investigating a robbery. He noted that the suspects fit the description of the man and woman who had committed the crime, except that the man did not have a beard, though he admitted that he sometimes wore one. Later that day the Los Angeles police arrested the two suspects, Malcolm Ricardo Collins, and his wife, Janet.

The evidence against the couple was scant, and the case rested heavily on the identification by the victim and the witness, John Bass. Unfortunately for the prosecution, neither proved to be a star on the witness stand. The victim could not identify Janet as the perpetrator and hadn't seen the driver at all. John Bass had not seen the perpetrator and said at the police lineup that he could not positively identify Malcolm Collins as the driver. And so, it seemed, the case was falling apart.

Enter the star witness, described in the California Supreme Court opinion only as "an instructor of mathematics at a state college." This witness testified that the fact that the defendants were "a Caucasian woman with a blond ponytail . . . [and] a Negro with a beard and mustache" who drove a partly yellow automobile was enough to convict the couple. To illustrate its point, the prosecution presented this table, quoted here verbatim from the supreme court decision:

Characteristic	Individual Probability
Partly yellow automobile	$\frac{1}{10}$
Man with mustache	$\frac{1}{4}$
Negro man with beard	$\frac{1}{10}$
Girl with ponytail	$\frac{1}{10}$
Girl with blond hair	$\frac{1}{3}$
Interracial couple in car	$\frac{1}{1,000}$

The math instructor called by the prosecution said that the product rule applies to this data. By multiplying all the probabilities, one concludes that the chances of a couple fitting all these distinctive characteristics are 1 in 12 million. Accordingly, he said, one could infer that the chances that the couple was innocent were 1 in 12 million. The prosecutor then pointed out that these individual probabilities were estimates and invited the jurors to supply their own guesses and then do the math. He himself, he said, believed they were conservative estimates, and the probability he came up with employing the factors he assigned was more like 1 in 1 billion. The jury bought it and convicted the couple.

What is wrong with this picture? For one thing, as we've seen, in order to find a compound probability by multiplying the component probabilities, the categories have to be independent, and in this case they clearly aren't. For example, the table quotes the chance of observing a "Negro man with beard" as 1 in 10 and a "man with mustache" as 1 in 4. But most men with a beard also have a mustache, so if you observe a "Negro man with beard," the chances are no longer 1 in 4 that the man you observe has a mustache—they are much higher. That issue can be remedied if you eliminate the category "Negro man with beard." Then the product of the probabilities falls to about 1 in 1 million.

There is another error in the analysis: the relevant probability is not the one stated above—the probability that a couple selected at random will match the suspects' description. Rather, the relevant probability is the chance that a couple matching all these characteristics is the guilty couple. The former might be 1 in 1 million. But as for the latter, the population of the area adjoining the one where the crime was committed was several million, so you might reasonably expect there to be 2 or 3 couples in the area who matched the description. In that case the probability that a couple who matched the description was guilty, based on this evidence alone (which is pretty much all the prosecution had), is only 1 in 2 or 3. Hardly beyond a reasonable doubt. For these reasons the supreme court overturned Collins's conviction.

The use of probability and statistics in modern courtrooms is still a controversial subject. In the Collins case the California Supreme Court derided what it called "trial by mathematics," but it left the door open to more "proper applications of mathematical techniques." In the ensuing years, courts rarely considered mathematical arguments, but even when attorneys and judges don't quote explicit probabilities or mathematical theorems, they do often employ this sort of reasoning, as do jurors when they weigh the evidence. Moreover, statistical arguments are becoming increasingly important because of the necessity of assessing DNA evidence. Unfortunately, with this increased importance has not come increased understanding on the part of attorneys, judges, or juries. As explained by Thomas Lyon, who teaches probability and the law at the University of Southern California, "Few students take a probability in law course, and few attorneys feel it has a place."[23] In law as in other realms, the understanding of randomness can reveal hidden layers of truth, but only to those who possess the tools to uncover them. In the next chapter we shall consider the story of the first man to study those tools systematically.

CHAPTER 3

Finding Your Way through
a Space of Possibilities

I N THE YEARS leading up to 1576, an oddly attired old man
could be found roving with a strange, irregular gait up and down
the streets of Rome, shouting occasionally to no one in particular
and being listened to by no one at all. He had once been celebrated
throughout Europe, a famous astrologer, physician to nobles of the
court, chair of medicine at the University of Pavia. He had created
enduring inventions, including a forerunner of the combination lock
and the universal joint, which is used in automobiles today. He had
published 131 books on a wide range of topics in philosophy, medicine, mathematics, and science. In 1576, however, he was a man
with a past but no future, living in obscurity and abject poverty. In
the late summer of that year he sat at his desk and wrote his final
words, an ode to his favorite son, his oldest, who had been executed
sixteen years earlier, at age twenty-six. The old man died on September 20, a few days shy of his seventy-fifth birthday. He had outlived
two of his three children; at his death his surviving son was employed
by the Inquisition as a professional torturer. That plum job was a
reward for having given evidence against his father.

Before his death, Gerolamo Cardano burned 170 unpublished
manuscripts.[1] Those sifting through his possessions found 111 that
survived. One, written decades earlier and, from the looks of it, often

41

revised, was a treatise of thirty-two short chapters. Titled *The Book on Games of Chance*, it was the first book ever written on the theory of randomness. People had been gambling and coping with other uncertainties for thousands of years. Can I make it across the desert before I die of thirst? Is it dangerous to remain under the cliff while the earth is shaking like this? Does that grin from the cave girl who likes to paint buffaloes on the sides of rocks mean she likes me? Yet until Cardano came along, no one had accomplished a reasoned analysis of the course that games or other uncertain processes take. Cardano's insight into how chance works came embodied in a principle we shall call the law of the sample space. The law of the sample space represented a new idea and a new methodology and has formed the basis of the mathematical description of uncertainty in all the centuries that followed. It is a simple methodology, a laws-of-chance analog of the idea of balancing a checkbook. Yet with this simple method we gain the ability to approach many problems systematically that would otherwise prove almost hopelessly confusing. To illustrate both the use and the power of the law, we shall consider a problem that although easily stated and requiring no advanced mathematics to solve, has probably stumped more people than any other in the history of randomness.

AS NEWSPAPER COLUMNS GO, *Parade* magazine's "Ask Marilyn" has to be considered a smashing success. Distributed in 350 newspapers and boasting a combined circulation of nearly 36 million, the question-and-answer column originated in 1986 and is still going strong. The questions can be as enlightening as the answers, an (unscientific) Gallup Poll of what is on Americans' minds. For instance:

When the stock market closes at the end of the day, why does everyone stand around smiling and clapping regardless of whether the stocks are up or down?

A friend is pregnant with twins that she knows are fraternal. What are the chances that at least one of the babies is a girl?

> When you drive by a dead skunk in the road, why does it take about 10 seconds before you smell it? Assume that you did not actually drive over the skunk.

Apparently Americans are a very practical people. The thing to note here is that each of the queries has a certain scientific or mathematical component to it, a characteristic of many of the questions answered in the column.

One might ask, especially if one knows a little something about mathematics and science, "Who is this guru Marilyn?" Well, Marilyn is Marilyn vos Savant, famous for being listed for years in the *Guinness World Records* Hall of Fame as the person with the world's highest recorded IQ (228). She is also famous for being married to Robert Jarvik, inventor of the Jarvik artificial heart. But sometimes famous people, despite their other accomplishments, are remembered for something they wished had never happened ("I did not have sexual relations with that woman"). That may be the case for Marilyn, who is most famous for her response to the following question, which appeared in her column one Sunday in September 1990 (I have altered the wording slightly):

> Suppose the contestants on a game show are given the choice of three doors: Behind one door is a car; behind the others, goats. After a contestant picks a door, the host, who knows what's behind all the doors, opens one of the unchosen doors, which reveals a goat. He then says to the contestant, "Do you want to switch to the other unopened door?" Is it to the contestant's advantage to make the switch?[2]

The question was inspired by the workings of the television game show *Let's Make a Deal*, which ran from 1963 to 1976 and in several incarnations from 1980 to 1991. The show's main draw was its handsome, amiable host, Monty Hall, and his provocatively clad assistant, Carol Merrill, Miss Azusa (California) of 1957.

It had to come as a surprise to the show's creators that after airing

43

4,500 episodes in nearly twenty-seven years, it was this question of mathematical probability that would be their principal legacy. This issue has immortalized both Marilyn and *Let's Make a Deal* because of the vehemence with which Marilyn vos Savant's readers responded to the column. After all, it appears to be a pretty silly question. Two doors are available—open one and you win; open the other and you lose—so it seems self-evident that whether you change your choice or not, your chances of winning are 50/50. What could be simpler? The thing is, Marilyn said in her column that it is better to switch.

Despite the public's much-heralded lethargy when it comes to mathematical issues, Marilyn's readers reacted as if she'd advocated ceding California back to Mexico. Her denial of the obvious brought her an avalanche of mail, 10,000 letters by her estimate.[3] If you ask the American people whether they agree that plants create the oxygen in the air, light travels faster than sound, or you cannot make radioactive milk safe by boiling it, you will get double-digit disagreement in each case (13 percent, 24 percent, and 35 percent, respectively).[4] But on this issue, Americans were united: 92 percent agreed Marilyn was wrong.

Many readers seemed to feel let down. How could a person they trusted on such a broad range of issues be confused by such a simple question? Was her mistake a symbol of the woeful ignorance of the American people? Almost 1,000 PhDs wrote in, many of them math professors, who seemed to be especially irate.[5] "You blew it," wrote a mathematician from George Mason University:

> Let me explain: If one door is shown to be a loser, that information changes the probability of either remaining choice— neither of which has any reason to be more likely—to ½. As a professional mathematician, I'm very concerned with the general public's lack of mathematical skills. Please help by confessing your error and, in the future, being more careful.

From Dickinson State University came this: "I am in shock that after being corrected by at least three mathematicians, you still do not see your mistake." From Georgetown: "How many irate mathematicians are needed to change your mind?" And someone from the U.S. Army Research Institute remarked, "If all those PhDs are wrong the country would be in serious trouble." Responses continued in such great numbers and for such a long time that after devoting quite a bit of column space to the issue, Marilyn decided she would no longer address it.

The army PhD who wrote in may have been correct that if all those PhDs were wrong, it would be a sign of trouble. But Marilyn *was* correct. When told of this, Paul Erdös, one of the leading mathematicians of the twentieth century, said, "That's impossible." Then, when presented with a formal mathematical proof of the correct answer, he still didn't believe it and grew angry. Only after a colleague arranged for a computer simulation in which Erdös watched hundreds of trials that came out 2 to 1 in favor of switching did Erdös concede he was wrong.[6]

How can something that seems so obvious be wrong? In the words of a Harvard professor who specializes in probability and statistics, "Our brains are just not wired to do probability problems very well."[7] The great American physicist Richard Feynman once told me never to think I understood a work in physics if all I had done was read someone else's derivation. The only way to really understand a theory, he said, is to derive it yourself (or perhaps end up disproving it!). For those of us who aren't Feynman, re-proving other people's work is a good way to end up untenured and plying our math skills as a checker at Home Depot. But the Monty Hall problem is one of those that can be solved without any specialized mathematical knowledge. You don't need calculus, geometry, algebra, or even amphetamines, which Erdös was reportedly fond of taking.[8] (As legend has it, once after quitting for a month, he remarked, "Before, when I looked at a piece of blank paper my mind was filled with ideas. Now all I see is a blank piece of paper.") All you need is a basic understanding of how probability works and the law of the sample space, that framework for analyzing chance situations that was first put on paper in the sixteenth century by Gerolamo Cardano.

GEROLAMO CARDANO was no rebel breaking forth from the intellectual milieu of sixteenth-century Europe. To Cardano a dog's howl portended the death of a loved one, and a few ravens croaking on the roof meant a grave illness was on its way. He believed as much as anyone else in fate, in luck, and in seeing your future in the alignment of planets and stars. Still, had he played poker, he wouldn't have been found drawing to an inside straight. For Cardano, gambling was second nature. His feeling for it was seated in his gut, not in his head, and so his understanding of the mathematical relationships among a game's possible random outcomes transcended his belief that owing to fate, any such insight is futile. Cardano's work also transcended the primitive state of mathematics in his day, for algebra and even arithmetic were yet in their stone age in the early sixteenth century, preceding even the invention of the equal sign.

History has much to say about Cardano, based on both his autobiography and the writings of some of his contemporaries. Some of the writings are contradictory, but one thing is certain: born in 1501, Gerolamo Cardano was not a child you'd have put your money on. His mother, Chiara, despised children, though—or perhaps *because*—she already had three boys. Short, stout, hot tempered, and promiscuous, she prepared a kind of sixteenth-century morning-after pill when she became pregnant with Gerolamo—a brew of wormwood, burned barleycorn, and tamarisk root. She drank it down in an attempt to abort the fetus. The brew sickened her, but the unborn Gerolamo was unfazed, perfectly content with whatever metabolites the concoction left in his mother's bloodstream. Her other attempts met with similar failure.

Chiara and Gerolamo's father, Fazio Cardano, were not married, but they often acted as if they were—they were known for their many loud quarrels. A month before Gerolamo's birth, Chiara left their home in Milan to live with her sister in Pavia, twenty miles to the south. Gerolamo emerged after three days of painful labor. One look at the infant and Chiara must have thought she would be rid of him after all. He was frail, and worse, lay silent. Chiara's midwife predicted he'd be dead within the hour. But if Chiara was thinking, *good riddance,* she was let down again, for the baby's wet nurse soaked him in a bath of warm wine, and Gerolamo revived. The infant's good health lasted only a few months, however. Then he, his nurse, and his three half brothers all came down with the plague. The Black Death, as the plague is sometimes called, is really three distinct diseases: bubonic, pneumonic, and septicemic plague. Cardano contracted bubonic, the most common, named for the buboes, the painful egg-size swellings of the lymph nodes that are one of the disease's prominent symptoms. Life expectancy, once buboes appeared, was about a week.

The Black Death had first entered Europe through a harbor in Messina in northeastern Sicily in 1347, carried by a Genoese fleet returning from the Orient.[9] The fleet was quickly quarantined, and the entire crew died aboard the ship—but the rats survived and scurried ashore, carrying both the bacteria and the fleas that would spread them. The ensuing outbreak killed half the city within two months and, eventually, between 25 percent and 50 percent of the population of Europe. Successive epidemics kept coming, tamping down the population of Europe for centuries. The year 1501 was a bad one for the plague in Italy. Gerolamo's nurse and brothers died. The lucky baby got away with nothing but disfigurement: warts on his nose, forehead, cheeks, and chin. He was destined to live nearly seventy-five years. Along the way there was plenty of disharmony and, in his early years, a good many beatings.

Gerolamo's father was a bit of an operator. A sometime pal of Leonardo da Vinci's, he was by profession a geometer, never a profession that brought in much cash. Fazio often had trouble making the rent, so he started a consulting business, providing the highborn with advice on law and medicine. That enterprise eventually thrived, aided by Fazio's claim that he was descended from a brother of a fellow named Goffredo Castiglioni of Milan, better known as Pope Celestine IV. When Gerolamo reached the age of five, his father brought him into the business—in a manner of speaking. That is, he strapped a pannier to his son's back, stuffed it with heavy legal and medical books, and began dragging the young boy to meetings with his patrons all over town. Gerolamo would later write that "from time to time as we walked the streets my father would command me to stop while he opened a book and, using my head as a table, read some long passage, prodding me the while with his foot to keep still if I wearied of the great weight."[10]

In 1516, Gerolamo decided his best opportunity lay in the field of medicine and announced that he wanted to leave his family's home in Milan and travel back to Pavia to study there. Fazio wanted him to study law, however, because then he would become eligible for an annual stipend of 100 crowns. After a huge family brawl, Fazio relented, but the question remained: without the stipend, how would Gerolamo support himself in Pavia? He began to save the money he earned reading horoscopes and tutoring pupils in geometry, alchemy, and astronomy. Somewhere along the way he noticed he had a talent for gambling, a talent that would bring him cash much faster than any of those other means.

For anyone interested in gambling in Cardano's day, every city was Las Vegas. On cards, dice, backgammon, even chess, wagers were made everywhere. Cardano classified these games according to two types: those that involved some strategy, or skill, and those that were governed by pure chance. In games like chess, Cardano risked being outplayed by some sixteenth-century Bobby Fischer. But when he bet on the fall of a couple of small cubes, his chances were as good as anyone else's. And yet in those games he did have an advantage, because he had developed a better understanding of the odds of winning in various situations than any of his opponents. And so for his entrée into the betting world, Cardano played the games of pure chance. Before long he had set aside over 1,000 crowns for his education—more than a decade's worth of the stipend his father wanted for him. In 1520 he registered as a student in Pavia. Soon after, he began to write down his theory of gambling.

. . .

LIVING WHEN HE DID, Cardano had the advantage of under-standing many things that had been Greek to the Greeks, and to the Romans, for the Hindus had taken the first large steps toward employing arithmetic as a powerful tool. It was in that milieu that positional notation in base ten developed, and became standard, around A.D. 700.[11] The Hindus also made great progress in the arith-metic of fractions—something crucial to the analysis of probabilities, since the chances of something occurring are always less than one. This Hindu knowledge was picked up by the Arabs and eventually brought to Europe. There the first abbreviations, *p* for "plus" and *m* for "minus," were used in the fifteenth century. The symbols + and − were introduced around the same time by the Germans, but only to indicate excess and deficient weights of chests. It gives one a feeling for some of the challenges Cardano faced to note that the equal sign did not yet exist, to be invented in 1557 by Robert Recorde of Oxford and Cambridge, who, inspired by geometry, remarked that no things could be more nearly alike than parallel lines and hence decided that such lines should denote equality. And the symbol ×, for multiplica-tion, attributable to an Anglican minister, didn't arrive on the scene until the seventeenth century.

Cardano's *Book on Games of Chance* covers card games, dice, backgammon, and astragali. It is not perfect. In its pages are reflected Cardano's character, his crazy ideas, his wild temper, the passion with which he approached every undertaking—and the turbulence of his life and times. It considers only processes—such as the toss of a die or the dealing of a playing card—in which one outcome is as likely as another. And some points Cardano gets wrong. Still, *The Book on Games of Chance* represents a beachhead, the first success in the human quest to understand the nature of uncertainty. And Car-dano's method of attacking questions of chance is startling both in its power and in its simplicity.

Not all the chapters of Cardano's book treat technical issues. For instance, chapter 26 is titled "Do Those Who Teach Well Also Play Well?" (he concludes, "It seems to be a different thing to know and to execute"). Chapter 29 is called "On the Character of Players"

("There are some who with many words drive both themselves and others from their proper senses"). These seem more "Dear Abby" than "Ask Marilyn." But then there is chapter 14, "On Combined Points" (on possibilities). There Cardano states what he calls "a general rule"—our law of the sample space.

The term *sample space* refers to the idea that the possible outcomes of a random process can be thought of as the points in a space. In simple cases the space might consist of just a few points, but in more complex situations it can be a continuum, just like the space we live in. Cardano didn't call it a space, however: the notion that a set of numbers could form a space was a century off, awaiting the genius of Descartes, his invention of coordinates, and his unification of algebra and geometry.

In modern language, Cardano's rule reads like this: *Suppose a random process has many equally likely outcomes, some favorable (that is, winning), some unfavorable (losing). Then the probability of obtaining a favorable outcome is equal to the proportion of outcomes that are favorable. The set of all possible outcomes is called the sample space.* In other words, if a die can land on any of six sides, those six outcomes form the sample space, and if you place a bet on, say, two of them, your chances of winning are 2 in 6.

A word on the assumption that all the outcomes are equally likely. Obviously that's not always true. The sample space for observing Oprah Winfrey's adult weight runs (historically) from 145 pounds to 237 pounds, and over time not all weight intervals have proved equally likely.[12] The complication that different possibilities have different probabilities can be accounted for by associating the proper odds with each possible outcome—that is, by careful accounting. But for now we'll look at examples in which all outcomes are equally probable, like those Cardano analyzed.

The potency of Cardano's rule goes hand in hand with certain subtleties. One lies in the meaning of the term *outcomes*. As late as the eighteenth century the famous French mathematician Jean Le Rond d'Alembert, author of several works on probability, misused the concept when he analyzed the toss of two coins.[13] The number of heads that turns up in those two tosses can be 0, 1, or 2. Since there are three outcomes, Alembert reasoned, the chances of each must be 1 in 3. But Alembert was mistaken.

One of the greatest deficiencies of Cardano's work was that he made no systematic analysis of the different ways in which a series of events, such as coin tosses, can turn out. As we shall see in the next chapter, no one did that until the following century. Still, a series of two coin tosses is simple enough that Cardano's methods are easily applied. The key is to realize that the possible outcomes of coin flipping are the data describing how the two coins land, not the total number of heads calculated *from* that data, as in Alembert's analysis. In other words, we should not consider 0, 1, or 2 heads as the possible outcomes but rather the sequences (heads, heads), (heads, tails), (tails, heads), and (tails, tails). These are the 4 possibilities that make up the sample space.

The next step, according to Cardano, is to sort through the outcomes, cataloguing the number of heads we can harvest from each. Only 1 of the 4 outcomes—(heads, heads)—yields 2 heads. Similarly, only (tails, tails) yields 0 heads. But if we desire 1 head, then 2 of the outcomes are favorable: (heads, tails) and (tails, heads). And so Cardano's method shows that Alembert was wrong: the chances are 25 percent for 0 or 2 heads but 50 percent for 1 head. Had Cardano laid his cash on 1 head at 3 to 1, he would have lost only half the time but tripled his money the other half, a great opportunity for a sixteenth-century kid trying to save up money for college—and still a great opportunity today if you can find anyone offering it.

A related problem often taught in elementary probability courses is the two-daughter problem, which is similar to one of the questions I quoted from the "Ask Marilyn" column. Suppose a mother is carrying fraternal twins and wants to know the odds of having two girls, a boy and a girl, and so on. Then the sample space consists of all the possible lists of the sexes of the children in their birth order: (girl, girl), (girl, boy), (boy, girl), and (boy, boy). It is the same as the space for the coin-toss problem except for the way we name the points: *heads* becomes *girl*, and *tails* becomes *boy*. Mathematicians have a fancy name for the situation in which one problem is another in disguise: they call it an isomorphism. When you find an isomorphism, it often means you've saved yourself a lot of work. In this case it means we can figure the chances that both children will be girls in exactly the same way we figured the chances of both tosses coming up heads in the coin-toss problem. And so without even doing the analysis, we know that the answer is the same: 25 percent. We can now answer the question asked in Marilyn's column: the chance that at least one of the babies will be a girl is the chance that both will be girls plus the chance that just one will be a girl—that is, 25 percent plus 50 percent, which is 75 percent.

In the two-daughter problem, an additional question is usually asked: What are the chances, *given that one of the children is a girl*, that both children will be girls? One might reason this way: since it is given that one of the children is a girl, there is only one child left to look at. The chance of that child's being a girl is 50 percent, so the probability that both children are girls is 50 percent.

That is not correct. Why? Although the statement of the problem says that one child is a girl, it doesn't say *which* one, and that changes things. If that sounds confusing, that's okay, because it provides a good illustration of the power of Cardano's method, which makes the reasoning clear.

The new information—one of the children is a girl—means that we are eliminating from consideration the possibility that both children are boys. And so, employing Cardano's approach, we eliminate the possible outcome (boy, boy) from the sample space. That leaves only 3 outcomes in the sample space: (girl, boy), (boy, girl), and (girl, girl). Of these, only (girl, girl) is the favorable outcome—that is, both children are daughters—so the chances that both children are girls is 1 in 3, or 33 percent. Now we can see why it matters that the statement of the problem didn't specify which child was a daughter. For instance, if the problem had asked for the chances of both children being girls *given that the first child is a girl*, then we would have eliminated both (boy, boy) and (boy, girl) from the sample space and the odds would have been 1 in 2, or 50 percent.

One has to give credit to Marilyn vos Savant, not only for attempting to raise public understanding of elementary probability but also for having the courage to continue to publish such questions even after her frustrating Monty Hall experience. We will end this discussion with another question taken from her column, this one from March 1996:

> My dad heard this story on the radio. At Duke University, two students had received A's in chemistry all semester. But on the night before the final exam, they were partying in another state and didn't get back to Duke until it was over. Their excuse to the professor was that they had a flat tire, and they asked if they could take a make-up test. The professor agreed, wrote out a test, and sent the two to separate rooms to take it. The first question (on one side of the paper) was worth five points. Then they flipped the paper over and found the second question, worth 95 points: "which tire was it?" What was the probability that both students would say the same thing? My dad and I think it's 1 in 16. Is that right?[14]

No, it is not: If the students were lying, the correct probability of their choosing the same answer is 1 in 4 (if you need help to see why, you can look at the notes at the back of this book).[15] And now that we're accustomed to decomposing a problem into lists of possibilities, we are ready to employ the law of the sample space to tackle the Monty Hall problem.

AS I SAID EARLIER, understanding the Monty Hall problem requires no mathematical training. But it does require some careful logical thought, so if you are reading this while watching *Simpsons* reruns, you might want to postpone one activity or the other. The good news is it goes on for only a few pages.

In the Monty Hall problem you are facing three doors: behind one door is something valuable, say a shiny red Maserati; behind the other two, an item of far less interest, say the complete works of Shakespeare in Serbian. You have chosen door 1. The sample space in this case is this list of three possible outcomes:

Maserati is behind door 1.
Maserati is behind door 2.
Maserati is behind door 3.

Each of these has a probability of 1 in 3. Since the assumption is that most people would prefer the Maserati, the first case is the winning case, and your chances of having guessed right are 1 in 3.

Now according to the problem, the next thing that happens is that the host, who knows what's behind all the doors, opens one you did not choose, revealing one of the sets of Shakespeare. In opening this door, the host has used what he knows to avoid revealing the Maserati, so this is *not* a completely random process. There are two cases to consider.

One is the case in which your initial choice was correct. Let's call that the Lucky Guess scenario. The host will now randomly open door 2 or door 3, and, if you choose to switch, instead of enjoying a fast, sexy ride, you'll be the owner of *Troilus and Cressida* in the Torlakian dialect. In the Lucky Guess scenario you are better off not switching—but the probability of landing in the Lucky Guess scenario is only 1 in 3.

The other case we must consider is that in which your initial choice was wrong. We'll call that the Wrong Guess scenario. The chances you guessed wrong are 2 out of 3, so the Wrong Guess scenario is twice as likely to occur as the Lucky Guess scenario. How does the Wrong Guess scenario differ from the Lucky Guess scenario? In the Wrong Guess scenario the Maserati is behind one of the doors you did not choose, and a copy of the Serbian Shakespeare is behind the other unchosen door. Unlike the Lucky Guess scenario, in this scenario the host does not randomly open an unchosen door. Since he does not want to reveal the Maserati, he *chooses* to open precisely the door that does *not* have the Maserati behind it. In other words, in the Wrong Guess scenario the host intervenes in what until now has been a random process. So the process is no longer random: the host uses his knowledge to bias the result, violating randomness by *guaranteeing* that if you switch your choice, you will get the fancy red car. Because of this intervention, if you find yourself in the Wrong Guess scenario, you will win if you switch and lose if you don't.

To summarize: if you are in the Lucky Guess scenario (probability 1 in 3), you'll win if you stick with your choice. If you are in the Wrong Guess scenario (probability 2 in 3), owing to the actions of the host, you will win if you switch your choice. And so your decision comes down to a guess: in which scenario do you find yourself? If you feel that ESP or fate has guided your initial choice, maybe you shouldn't switch. But unless you can bend silver spoons into pretzels with your brain waves, the odds are 2 to 1 that you are in the Wrong Guess scenario, and so it is better to switch. Statistics from the television program bear this out: those who found themselves in the situation described in the problem and switched their choice won about twice as often as those who did not.

The Monty Hall problem is hard to grasp because unless you think about it carefully, the role of the host, like that of your mother, goes unappreciated. But the host is fixing the game. The host's role can be made obvious if we suppose that instead of 3 doors, there were 100. You still choose door 1, but now you have a probability of 1 in 100 of being right. Meanwhile the chance of the Maserati's being behind one of the other doors is 99 in 100. As before, the host opens all but one of the doors that you did not pick, being sure not to open the door hiding the Maserati if it is one of them. After he is done, the chances are still 1 in 100 that the Maserati was behind the door you chose and still 99 in 100 that it was behind one of the other doors. But now, thanks to the intervention of the host, there is only one door left representing all 99 of those other doors, and so the probability that the Maserati is behind that remaining door is 99 out of 100!

Had the Monty Hall problem been around in Cardano's day, would he have been a Marilyn vos Savant or a Paul Erdös? The law of the sample space handles the problem nicely, but we have no way of knowing for sure, for the earliest known statement of the problem (under a different name) didn't occur until 1959, in an article by Martin Gardner in *Scientific American*.[16] Gardner called it "a wonderfully confusing little problem" and noted that "in no other branch of mathematics is it so easy for experts to blunder as in probability theory." Of course, to a mathematician a blunder is an issue of embarrassment, but to a gambler it is an issue of livelihood. And so it is fitting that when it came to the first systematic theory of probability, it took Cardano, the gambler, to figure things out.

ONE DAY while Cardano was in his teens, one of his friends died suddenly. After a few months, Cardano noticed, his friend's name was no longer mentioned by anyone. This saddened him and left a deep impression. How does one overcome the fact that life is transitory? He decided that the only way was to leave something behind—heirs or lasting works of some kind or both. In his autobiography, Cardano describes developing "an unshakable ambition" to leave his mark on the world.[17]

After obtaining his medical degree, Cardano returned to Milan, seeking employment. While in college he had written a paper, "On the Differing Opinions of Physicians," that essentially called the medical establishment a bunch of quacks. The Milan College of Physicians now returned the favor, refusing to admit him. That meant he could not practice in Milan. And so, using money he had saved from his tutoring and gambling, Cardano bought a tiny house to the east, in the town of Piove di Sacco. He expected to do good business there because disease was rife in the town and it had no physician. But his market research had a fatal flaw: the town had no doctor because the populace preferred to be treated by sorcerers and priests. After years of intense work and study, Cardano found himself with little income but a lot of spare time on his hands. It proved a lucky break, for he seized the opportunity and began to write books. One of them was *The Book on Games of Chance*.

In 1532, after five years in Sacco, Cardano moved back to Milan, hoping to have his work published and once again applying for membership in the College of Physicians. On both fronts he was roundly rejected. "In those days," he wrote, "I was sickened so to the heart that I would visit diviners and wizards so that some solution might be found to my manifold troubles."[18] One wizard suggested he shield himself from moon rays. Another that, on waking, he sneeze three times and knock on wood. Cardano followed all their prescriptions, but none changed his bad fortune. And so, hooded, he took to sneaking from building to building at night, surreptitiously treating patients who either couldn't afford the fees of sanctioned doctors or else didn't improve in their care. To supplement the income he earned from that endeavor, he wrote in his autobiography, he was "forced to the dice again so that I could support my wife; and here my knowledge defeated fortune, and we were able to buy food and live, though our lodgings were desolate."[19] As for *The Book on Games of Chance*, though he would revise and improve the manuscript repeatedly in the years to come, he never again sought to have it published, perhaps because he realized it wasn't a good idea to teach anyone to gamble as well as he could.

Cardano eventually achieved his goals in life, obtaining both heirs and fame—and a good deal of fortune to boot. The fortune began to accrue when he published a book based on his old college paper, altering the title from the somewhat academic "On the Differing Opinions of Physicians" to the zinger *On the Bad Practice of Medicine in Common Use*. The book was a hit. And then, when one of his secret patients, a well-known prior of the Augustinian order of friars, suddenly (and in all likelihood by chance) improved and attributed his recovery to Cardano's care, Cardano's fame as a physician took off on an upward spiral that reached such heights the College of Physicians felt compelled not only to grant him membership but also to make him its rector. Meanwhile he was publishing more books, and they did well, especially one for the general public called *The Practice of Arithmetic*. A few years later he published a more technical book, called the *Ars magna*, or *The Great Art*, a treatise on algebra in which he gave the first clear picture of negative numbers and a famous analysis of certain algebraic equations. When he reached his early fifties, in the mid-1550s, Cardano was at his peak, chairman of medicine at the University of Pavia and a wealthy man.

His good fortune didn't last. To a large extent what brought Cardano down was the other part of his legacy—his children. When she was sixteen, his daughter Chiara (named after his mother) seduced his older son, Giovanni, and become pregnant. She had a successful abortion, but it left her infertile. That suited her just fine, for she was boldly promiscuous, even after her marriage, and contracted syphilis. Giovanni went on to become a doctor but was soon more famous as a petty criminal, so famous he was blackmailed into marriage by a family of gold diggers who had proof that he had murdered, by poison, a minor city official. Meanwhile Aldo, Cardano's younger son who as a child had engaged in the torture of animals, turned that passion into work as a freelance torturer for the Inquisition. And like Giovanni, he moonlighted as a crook.

A few years after his marriage Giovanni gave one of his servants a mysterious mixture to incorporate into a cake for Giovanni's wife. When she keeled over after enjoying her dessert, the authorities put two and two together. Despite Gerolamo's spending a fortune on lawyers, his attempts to pull strings, and his testimony on his son's behalf, young Giovanni was executed in prison a short while later. The drain on Cardano's funds and reputation made him vulnerable to his old enemies. The senate in Milan expunged his name from the list of those allowed to lecture, and accusing him of sodomy and incest, had him exiled from the province. When Cardano left Milan at the end of 1563, he wrote in his autobiography, he was "reduced once more to rags, my fortune gone, my income ceased, my rents withheld, my books impounded."[20] By that time his mind was going too, and he was given to periods of incoherence. As the final blow, a self-taught mathematician named Niccolò Tartaglia, angry because in *Ars magna* Cardano had revealed Tartaglia's secret method of solving certain equations, coaxed Aldo into giving evidence against his father in exchange for an official appointment as public torturer and executioner for the city of Bologna. Cardano was jailed briefly, then quietly lived out his last few years in Rome. *The Book on Games of Chance* was finally published in 1663, over 100 years after young Cardano had first put the words to paper. By then his methods of analysis had been reproduced and surpassed.

CHAPTER 4

Tracking the Pathways
to Success

I F A GAMBLER of Cardano's day had understood Cardano's mathematical work on chance, he could have made a tidy profit betting against less sophisticated players. Today, with what he had to offer, Cardano could have achieved both fame and fortune writing books like *The Idiot's Guide to Casting Dice with Suckers*. But in his own time, Cardano's work made no big splash, and his *Book on Games of Chance* remained unpublished until long after his death. Why did Cardano's work have so little impact? As we've said, one hindrance to those who preceded him was the lack of a good system of algebraic notation. That system in Cardano's day was improving but was still in its infancy. Another roadblock, however, had yet to be removed: Cardano worked at a time when mystical incantation was more valued than mathematical calculation. If people did not look for the order in nature and did not develop numerical descriptions of events, then a theory of the effect of randomness on those events was bound to go unappreciated. As it turned out, had Cardano lived just a few decades later, both his work and its reception might have been far different, for the decades after his death saw the unfolding of historic changes in European thought and belief, a transformation that has traditionally been dubbed the scientific revolution.

The scientific revolution was a revolt against a way of thinking that was prevalent as Europe emerged from the Middle Ages, an era in which people's beliefs about the way the world worked were not scrutinized in any systematic manner. Merchants in one town stole the clothes off a hanged man because they believed it would help their sales of beer. Parishioners in another believed illness could be cured by chanting sacrilegious prayers as they marched naked around their church altar.[1] One trader even believed that relieving himself in the "wrong" toilet would bring bad fortune. Actually he was a bond trader who confessed his secret to a CNN reporter in 2003.[2] Yes, some people still adhere to superstitions today, but at least today, for those who are interested, we have the intellectual tools to prove or disprove the efficacy of such actions. But if Cardano's contemporaries, say, won at dice, rather than analyzing their experience mathematically, they would say a prayer of thanks or refuse to wash their lucky socks. Cardano himself believed that streaks of losses occur because "fortune is averse" and that one way to improve your results is to give the dice a good hard throw. If a lucky 7 is all in the wrist, why stoop to mathematics?

The moment that is often considered the turning point for the scientific revolution came in 1583, just seven years after Cardano's death. That is when a young student at the University of Pisa sat in a cathedral and, according to legend, rather than listening to the services, stared at something he found far more intriguing: the swinging of a large hanging lamp. Using his pulse as a timer, Galileo Galilei noticed that the lamp seemed to take the same amount of time to swing through a wide arc as it did to swing through a narrow one. That observation suggested to him a law: the time required by a pendulum to perform a swing is independent of the amplitude of the swing. Galileo's was a precise and practical observation, and although simple, it signified a new approach to the description of physical phenomena: the idea that science must focus on experience and experimentation—how nature operates—rather than on what intuition dictates or our minds find appealing. And most of all, it must be done with mathematics.

Galileo employed his scientific skills to write a short piece on gambling, "Thoughts about Dice Games." The work was produced at the behest of his patron, the grand duke of Tuscany. The problem that bothered the grand duke was this: when you throw three dice, why does the number 10 appear more frequently than the number 9? The excess of 10s is only about 8 percent, and neither 10 nor 9 comes up very often, so the fact that the grand duke played enough to notice the small difference means he probably needed a good twelve-step program more than he needed Galileo. For whatever reason, Galileo was not keen to work on the problem and grumbled about it. But like any consultant who wants to stay employed, he kept his grumbling low-key and did his job.

If you throw a single die, the chances of any number in particular coming up are 1 in 6. But if you throw two dice, the chances of different totals are no longer equal. For example, there is a 1 in 36 chance of the dice totaling 2 but twice that chance of their totaling 3. The reason is that a total of 2 can be obtained in only 1 way, by tossing two 1s, but a total of 3 can be obtained in 2 ways, by tossing a 1 and then a 2 or a 2 and then a 1. That brings us to the next big step in understanding random processes, which is the subject of this chapter: the development of systematic methods for analyzing the number of ways in which events can happen.

THE KEY TO UNDERSTANDING the grand duke's confusion is to approach the problem as if you were a Talmudic scholar: rather than attempting to explain why 10 comes up more frequently than 9, we ask, *why shouldn't 10 come up more frequently than 9?* It turns out there is a tempting reason to believe that the dice should sum to 10 and 9 with equal frequency: both 10 and 9 can be constructed in 6 ways from the throw of three dice. For 9 we can write those ways as (621), (531), (522), (441), (432), and (333). For 10 they are (631), (622), (541), (532), (442), and (433). According to Cardano's law of the sample space, the probability of obtaining a favorable outcome is equal to the proportion of outcomes that are favorable. A sum of

9 and 10 can be constructed in the same number of ways. So why is one more probable than the other?

The reason is that, as I've said, the law of the sample space in its original form applies only to outcomes that are equally probable, and the combinations listed above are not. For instance, the outcome (631)—that is, throwing a 6, a 3, and a 1—is 6 times more likely than the outcome (333) because although there is only 1 way you can throw three 3s, there are 6 ways you can throw a 6, a 3, and a 1: you can throw a 6 first, then a 3, and then a 1, or you can throw a 1 first, then a 3, then a 6, and so on. Let's represent an outcome in which we are keeping track of the order of throws by a triplet of numbers separated by commas. Then the short way of saying what we just said is that the outcome (631) consists of the possibilities (1,3,6), (1,6,3), (3,1,6), (3,6,1), (6,1,3), and (6,3,1), whereas the outcome (333) consists only of (3,3,3). Once we've made this decomposition, we can see that the outcomes are equally probable and we can apply the law. Since there are 27 ways of rolling a 10 with three dice but only 25 ways to get a total of 9, Galileo concluded that with three dice, rolling a 10 was 27/25, or about 1.08, times more likely.

In solving the problem, Galileo implicitly employed our next important principle: *The chances of an event depend on the number of ways in which it can occur.* That is not a surprising statement. The surprise is just how large that effect is—and how difficult it can be to calculate. For example, suppose you give a 10-question true-or-false quiz to your class of 25 sixth-graders. Let's do an accounting of the results a particular student might achieve: she could answer all questions correctly; she could miss 1 question—that can happen in 10 ways because there are 10 questions she could miss; she could miss a pair of questions—that can happen in 45 ways because there are 45 distinct pairs of questions; and so on. As a result, on average in a collection of students who are randomly guessing, for every student scoring 100 percent, you'll find about 10 scoring 90 percent and 45 scoring 80 percent. The chances of getting a grade near 50 percent are of course higher still, but in a class of 25 the probability that at least one student will get a B (80 percent) or better if all the students are guessing is about 75 percent. So if you are a veteran teacher, it is likely that among all the students over the years who have shown up unprepared and more or less guessed at your quizzes, some were rewarded with an A or a B.

A few years ago Canadian lottery officials learned the importance of careful counting the hard way when they decided to give back some unclaimed prize money that had accumulated.[3] They purchased 500 automobiles as bonus prizes and programmed a computer to determine the winners by randomly selecting 500 numbers from their list of 2.4 million subscriber numbers. The officials published the unsorted list of 500 winning numbers, promising an automobile for each number listed. To their embarrassment, one individual claimed (rightly) that he had won two cars. The officials were flabbergasted—with over 2 million numbers to choose from, how could the computer have randomly chosen the same number twice? Was there a fault in their program?

The counting problem the lottery officials ran into is equivalent to a problem called the birthday problem: how many people must a group contain in order for there to be a better than even chance that two members of the group will share the same birthday (assuming all birth dates are equally probable)? Most people think the answer is half the number of days in a year, or about 183. But that is the correct answer to a different question: how many people do you need to have at a party for there to be a better than even chance that one of them will share *your* birthday? If there is no restriction on *which* two people will share a birthday, the fact that there are many possible pairs of individuals who might have shared birthdays changes the answer drastically. In fact, the answer is astonishingly low: just 23. When pulling from a pool of 2.4 million, as in the case of the Canadian lottery, it takes many more than 500 numbers to have an even chance of a repeat. But still that possibility should not have been ignored. The chances of a match come out, in fact, to about 5 percent. Not huge, but it could have been accounted for by having the computer cross each number off the list as it was chosen. For the record, the Canadian lottery requested the lucky fellow to forgo the second car, but he refused.

Another lottery mystery that raised many eyebrows occurred in Germany on June 21, 1995.[4] The freak event happened in a lottery called Lotto 6/49, which means that the winning six numbers are drawn from the numbers 1 to 49. On the day in question the winning numbers were 15-25-27-30-42-48. The very same sequence had been drawn previously, on December 20, 1986. It was the first time in 3,016 drawings that a winning sequence had been repeated. What were the chances of that? Not as bad as you'd think. When you do the math, the chance of a repeat at some point over the years comes out to around 28 percent.

Since in a random process the number of ways in which an outcome can occur is a key to determining how probable it is, the key question is, how do you calculate the number of ways in which something can occur? Galileo seems to have missed the significance of that question. He did not carry his work on randomness beyond that problem of dice and said in the first paragraph of his work that he was writing about dice only because he had been "ordered" to do so.[5] In 1633, as his reward for promoting a new approach to science, Galileo was condemned by the Inquisition. But science and theology had parted ways for good; scientists now analyzing how? were unburdened by the theologians' issue of why? Soon a scholar from a new generation, schooled since his youth on Galileo's philosophy of science, would take the analysis of contingency counting to new heights, reaching a level of understanding without which most of today's science could not be conducted.

WITH THE BLOSSOMING of the scientific revolution the frontiers of randomness moved from Italy to France, where a new breed of scientist, rebelling against Aristotle and following Galileo, developed it further and deeper than had either Cardano or Galileo. This time the importance of the new work would be recognized, and it would make waves all over Europe. Though the new ideas would again be developed in the context of gambling, the first of this new breed was more a mathematician turned gambler than, like Cardano, a gambler turned mathematician. His name was Blaise Pascal.

Pascal was born in June 1623 in Clermont-Ferrand, a little more than 250 miles south of Paris. Realizing his son's brilliance, and having moved to Paris, Blaise's father introduced him at age thirteen to a newly founded discussion group there that insiders called the Académie Mersenne after the black-robed friar who had founded it. Mersenne's society included the famed philosopher-mathematician René Descartes and the amateur mathematics genius Pierre de Fermat. The strange mix of brilliant thinkers and large egos, with Mersenne present to stir the pot, must have had a great influence on the teenage Blaise, who developed personal ties to both Fermat and Descartes and picked up a deep grounding in the new scientific method. "Let all the disciples of Aristotle . . . ," he would write, "recognize that experiment is the true master who must be followed in Physics."[6]

But how did a bookish and stodgy fellow of pious beliefs become involved with issues of the urban gambling scene? On and off Pascal experienced stomach pains, had difficulty swallowing and keeping food down, and suffered from debilitating weakness, severe headaches, bouts of sweating, and partial paralysis of the legs. He stoically followed the advice of his physicians, which involved bleedings, purgings, and the consumption of asses' milk and other "disgusting" potions that he could barely keep from vomiting—a "veritable torture," according to his sister Gilberte.[7] Pascal had by then left Paris, but in the summer of 1647, aged twenty-four and growing desperate, he moved back with his sister Jacqueline in search of better medical care. There his new bevy of doctors offered the state-of-the-art advice that Pascal "ought to give up all continued mental labor, and should seek as much as possible all opportunities to divert himself."[8] And so Pascal taught himself to kick back and relax and began to spend time in the company of other young men of leisure. Then, in 1651, Blaise's father died, and suddenly Pascal was a twenty-something with an inheritance. He put the cash to good use, at least in the sense of his doctors' orders. Biographers call the years from 1651 to 1654 Pascal's "worldly period." His sister Gilberte called it "the time of his life that was worst employed."[9] Though he put some effort into self-promotion, his scientific research went almost nowhere, but for the record, his health was the best it had ever been.

Often in history the study of the random has been aided by an event that was itself random. Pascal's work represents such an occasion, for it was his abandonment of study that led him to the study of chance. It all began when one of his partying pals introduced him to a forty-five-year-old snob named Antoine Gombaud. Gombaud, a nobleman whose title was chevalier de Méré, regarded himself as a master of flirtation, and judging by his catalog of romantic entanglements, he was. But de Méré was also an expert gambler who liked the stakes high and won often enough that some suspected him of cheating. And when he stumbled on a little gambling quandary, he turned to Pascal for help. With that, de Méré initiated an investigation that would bring to an end Pascal's scientific dry spell, cement de Méré's own place in the history of ideas, and solve the problem left open by Galileo's work on the grand duke's dice-tossing question.

The year was 1654. The question de Méré brought to Pascal was called the problem of points: Suppose you and another player are playing a game in which you both have equal chances and the first player to earn a certain number of points wins. The game is interrupted with one player in the lead. What is the fairest way to divide the pot? The solution, de Méré noted, should reflect each player's chance of victory given the score that prevails when the game is interrupted. But how do you calculate that?

Pascal realized that whatever the answer, the methods needed to calculate it were yet unknown, and those methods, whatever they were, could have important implications in any type of competitive situation. And yet, as often happens in theoretical research, Pascal found himself unsure of, and even confused about, his plan of attack. He decided he needed a collaborator, or at least another mathematician with whom he could discuss his ideas. Marin Mersenne, the great communicator, had died a few years earlier, but Pascal was still wired into the Académie Mersenne network. And so in 1654 began one of the great correspondences in the history of mathematics, between Pascal and Pierre de Fermat.

In 1654, Fermat held a high position in the Tournelle, or criminal court, in Toulouse. When the court was in session, a finely robed Fermat might be found condemning errant functionaries to be burned at the stake. But when the court was not in session, he would turn his analytic skills to the gentler pursuit of mathematics. He may have been an amateur, but Pierre de Fermat is usually considered the greatest amateur mathematician of all times.

Fermat had not gained his high position through any particular ambition or accomplishment. He achieved it the old-fashioned way, by moving up steadily as his superiors dropped dead of the plague. In fact, when Pascal's letter arrived, Fermat himself was recovering from a bout of the disease. He had even been reported dead, by his friend Bernard Medon. When Fermat didn't die, an embarrassed but presumably happy Medon retracted his announcement, but there is no doubt that Fermat had been on the brink. As it turned out, though twenty-two years Pascal's senior, Fermat would outlive his newfound correspondent by several years.

As we'll see, the problem of points comes up in any area of life in which two entities compete. In their letters, Pascal and Fermat each developed his own approach and solved several versions of the problem. But it was Pascal's method that proved simpler—even beautiful—and yet is general enough to be applied to many problems we encounter in our everyday experience. Because the problem of points first arose in a betting situation, I'll illustrate the problem with an example from the world of sports. In 1996 the Atlanta Braves beat the New York Yankees in the first 2 games of the baseball World Series, in which the first team to win 4 games is crowned champion. The fact that the Braves won the first 2 games didn't necessarily mean they were the superior team. Still, it could be taken as a sign that they were indeed better. Nevertheless, for our current purposes we will stick to the assumption that either team was equally likely to win each game and that the first 2 games just happened to go to the Braves.

Given that assumption, what would have been fair odds for a bet on the Yankees—that is, what was the chance of a Yankee comeback? To calculate it, we count all the ways in which the Yankees could have won and compare that to the number of ways in which they could have lost. Two games of the series had been played, so there were 5 possible games yet to play. And since each of those games had 2 possible outcomes—a Yankee win (Y) or a Braves win (B)—there were 2^5, or 32, possible outcomes. For instance, the Yankees could have won 3, then lost 2: YYYBB; or they could have alternated victories: YBYBY. (In the latter case, since the Braves would have won 4 games with the 6th game, the last game would never have been played, but we'll get to that in a minute.) The probability that the Yankees would come back to win the series was equal to the number of sequences in which they would win at least 4 games divided by the total number of sequences, 32; the chance that the Braves would win was equal to the number of sequences in which they would win at least 2 more games also divided by 32.

This calculation may seem odd, because as I mentioned, it includes scenarios (such as YBYBY) in which the teams keep playing even after the Braves have won the required 4 games. The teams would certainly not play a 7th game once the Braves had won 4. But mathematics is independent of human whim, and whether or not the players play the games does not affect the fact that such sequences exist. For example, suppose you're playing a coin-toss game in which you win if at any time heads come up. There are 2^2, or 4, possible two-toss sequences: HT, HH, TH, and TT. In the first two of these, you would not bother tossing the coin again because you would already have won. Still, your chances of winning are 3 in 4 because 3 of the 4 complete sequences include an H.

So in order to calculate the Yankees' and the Braves' chances of victory, we simply make an accounting of the possible 5-game sequences for the remainder of the series. First, the Yankees would have been victorious if they had won 4 of the 5 possible remaining games. That could have happened in 1 of 5 ways: BYYYY, YBYYY, YYBYY, YYYBY, or YYYYB. Alternatively, the Yankees would have triumphed if they had won all 5 of the remaining games, which could have happened in only 1 way: YYYYY. Now for the Braves: they would have become champions if the Yankees had won only 3 games, which could have happened in 10 ways (BBYYY, BYBYY, and so on), or if the Yankees had won only 2 games (which again could have happened in 10 ways), or if the Yankees had won only 1 game (which could have happened in 5 ways), or if they had won none (which could have happened in only 1 way). Adding these possible outcomes together, we find that the chance of a Yankees victory was 6 in 32, or about 19 percent, versus 26 in 32, or about 81 percent for the Braves. According to Pascal and Fermat, if the series had abruptly been terminated, that's how they should have split the bonus pot, and those are the odds that should have been set if a bet was to be made after the first 2 games. For the record, the Yankees did come back to win the next 4 games, and they were crowned champion.

The same reasoning could also be applied to the start of the series—that is, before any game has been played. If the two teams have equal chances of winning each game, you will find, of course, that they have an equal chance of winning the series. But similar reasoning works if they don't have an equal chance, except that the simple accounting I just employed would have to be altered slightly: each outcome would have to be weighted by a factor describing its relative probability. If you do that and analyze the situation at the start of the series, you will discover that in a 7-game series there is a sizable chance that the inferior team will be crowned champion. For instance, if one team is good enough to warrant beating another in 55 percent of its games, the weaker team will nevertheless win a 7-game series about 4 times out of 10. And if the superior team could be expected to beat its opponent, on average, 2 out of each 3 times they meet, the inferior team will still win a 7-game series about once every 5 matchups. There is really no way for sports leagues to change this. In the lopsided ⅔-probability case, for example, you'd have to

70

play a series consisting of at minimum the best of 23 games to determine the winner with what is called statistical significance, meaning the weaker team would be crowned champion 5 percent or less of the time (see chapter 5). And in the case of one team's having only a 55–45 edge, the shortest statistically significant "world series" would be the best of 269 games, a tedious endeavor indeed! So sports playoff series can be fun and exciting, but being crowned "world champion" is not a very reliable indication that a team is actually the best one.

As I said, the same reasoning applies to more than games, gambling, and sports. For example, it shows that if two companies compete head-to-head or two employees within a company compete, though there may be a winner and a loser each quarter or each year, to get a reliable answer regarding which company or which employee is superior by simply tallying who beats whom, you'd have to make the comparison over decades or centuries. If, for instance, employee A is truly superior and would in the long run win a performance comparison with employee B on 60 out of 100 occasions, in a simple best-of-5 series of comparisons the weaker employee will still win almost one-third of the time. It is dangerous to judge ability by short-term results.

The counting in all these problems has been simple enough to carry out without much effort. But when the numbers are higher, the counting becomes difficult. Consider, for example, this problem: You are arranging a wedding reception for 100 guests, and each table seats 10. You can't sit your cousin Rod with your friend Amy because eight years ago they had an affair and she dumped him. On the other hand, both Amy and Leticia want to sit next to your buff cousin Bobby, and your aunt Ruth had better be at a table out of earshot or the dueling flirtations will be gossip fodder for holiday dinners for the next five years. You carefully consider the possibilities. Take just the first table. How many ways are there to choose 10 people from a group of 100? That's the same question as, in how many ways can you apportion 10 investments among 100 mutual funds or 10 germanium atoms among 100 locations in a silicon crystal? It's the type of problem that comes up repeatedly in the theory of randomness, and not only in the problem of points. But with larger numbers it is tedious or impossible to count the possibilities by listing them explicitly. That was Pascal's real accomplishment: a generally applicable and systematic approach to counting that allows you to calculate the answer from a formula or read it off a chart. It is based on a curious arrangement of numbers in the shape of a triangle.

THE COMPUTATIONAL METHOD at the heart of Pascal's work was actually discovered by a Chinese mathematician named Jia Xian around 1050, published by another Chinese mathematician, Zhu Shijie, in 1303, discussed in a work by Cardano in 1570, and plugged into the greater whole of probability theory by Pascal, who ended up getting most of the credit.[10] But the prior work didn't bother Pascal. "Let no one say I have said nothing new," Pascal argued in his autobiography. "The arrangement of the subject is new. When we play tennis, we both play with the same ball, but one of us places it better."[11] The graphic invention employed by Pascal, given below, is thus called Pascal's triangle. In the figure, I have truncated Pascal's triangle at the tenth row, but it can be continued downward indefinitely. In fact, it is easy to continue the triangle, for with the exception of the 1 at the apex, each number is the sum of the number in the line

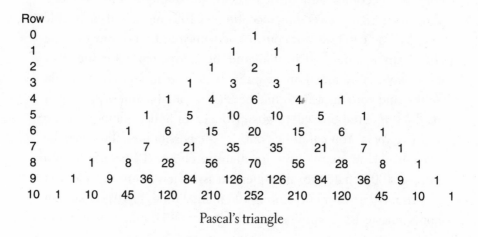

Row																				
0										1										
1									1		1									
2								1		2		1								
3							1		3		3		1							
4						1		4		6		4		1						
5					1		5		10		10		5		1					
6				1		6		15		20		15		6		1				
7			1		7		21		35		35		21		7		1			
8		1		8		28		56		70		56		28		8		1		
9	1		9		36		84		126		126		84		36		9		1	
10	1	10		45		120		210		252		210		120		45		10		1

Pascal's triangle

above it to the left and the number in the line above it to the right (add a 0 if there is no number in the line above it to the left or to the right).

Pascal's triangle is useful any time you need to know the number of ways in which you can choose some number of objects from a collection that has an equal or greater number. Here is how it works in the case of the wedding guests: To find the number of distinct seatings of 10 you can form from a group of 100 guests, you would start by looking down the numbers to the left of the triangle until you found the row labeled 100. The triangle I supplied does not go down that far, but for now let's pretend it does. The first number in row 100 tells you the number of ways you can choose 0 guests from a group of 100. There is just 1 way, of course: you simply don't choose anyone. That is true no matter how many total guests you are choosing from, which is why the first number in every row is a 1. The second number in row 100 tells you the number of ways you can choose 1 guest from the group of 100. There are 100 ways to do that: you can choose just guest number 1, or just guest number 2, and so on. That reasoning applies to every row, and so the second number in each row is simply the number of that row. The third number in each row represents the number of distinct groups of 2 you can form, and so on. The number we seek—the number of distinct arrangements of 10 you can form— is therefore the eleventh number in the row. Even if I had extended the triangle to include 100 rows, that number would be far too large to put on the page. In fact, when some wedding guest inevitably complains about the seating arrangements, you might point out how long it would have taken you to consider every possibility: assuming you spent one second considering each one, it would come to roughly 10,000 billion years. The unhappy guest will assume, of course, that you are being histrionic.

In order for us to use Pascal's triangle, let's say for now that your guest list consists of just 10 guests. Then the relevant row is the one at the bottom of the triangle I provided, labeled 10. The numbers in that row represent the number of distinct tables of 0, 1, 2, and so on, that can be formed from a collection of 10 people. You may recognize these numbers from the sixth-grade quiz example—the number of ways in which a student can get a given number of problems wrong on a 10-question true-or-false test is the same as the number of ways in which you can choose guests from a group of 10. That is one of the reasons for the power of Pascal's triangle: the same mathematics can be applied to many different situations. For the Yankees-Braves World Series example, in which we tediously counted all the possibilities for the remaining 5 games, we can now read the number of ways in which the Yankees can win 0, 1, 2, 3, 4, or 5 games directly from row 5 of the triangle:

$$1 \quad 5 \quad 10 \quad 10 \quad 5 \quad 1$$

We can see at a glance that the Yankees' chance of winning 2 games (10 ways) was twice as high as their chance of winning 1 game (5 ways).

Once you learn the method, applications of Pascal's triangle crop up everywhere. A friend of mine once worked for a start-up computer-games company. She would often relate how, although the marketing director conceded that small focus groups were suited for "qualitative conclusions only," she nevertheless sometimes reported an "overwhelming" 4-to-2 or 5-to-1 agreement among the members of the group as if it were meaningful. But suppose you hold a focus group in which 6 people will examine and comment on a new product you are developing. Suppose that in actuality the product appeals to half the population. How accurately will this preference be reflected in your focus group? Now the relevant line of the triangle is the one labeled 6, representing the number of possible subgroups of 0, 1, 2, 3, 4, 5, or 6 whose members might like (or dislike) your product:

$$1 \quad 6 \quad 15 \quad 20 \quad 15 \quad 6 \quad 1$$

From these numbers we see that there are 20 ways in which the group members could split 50/50, accurately reflecting the views of the populace at large. But there are also $1 + 6 + 15 + 15 + 6 + 1 = 44$

ways in which you might find an unrepresentative consensus, either for or against. So if you are not careful, the chances of being misled are 44 out of 64, or about two-thirds. This example does not prove that if agreement is achieved, it is random. But neither should you assume that it is significant.

Pascal and Fermat's analysis proved to be a big first step in a coherent mathematical theory of randomness. The final letter of their famous exchange is dated October 27, 1654. A few weeks later Pascal sat in a trance for two hours. Some call that trance a mystical experience. Others lament that he had finally blasted off from planet Sanity. However you describe it, Pascal emerged from the event a transformed man. It was a transformation that would lead him to make one more fundamental contribution to the concept of randomness.

IN 1662, a few days after Pascal died, a servant noticed a curious bulge in one of Pascal's jackets. The servant pulled open the lining to find hidden within it folded sheets of parchment and paper. Pascal had apparently carried them with him every day for the last eight years of his life. Scribbled on the sheets, in his handwriting, was a series of isolated words and phrases dated November 23, 1654. The writings were an emotional account of the trance, in which he described how God had come to him and in the space of two hours delivered him from his corrupt ways.

Following that revelation, Pascal had dropped most of his friends, calling them "horrible attachments."[12] He sold his carriage, his horses, his furniture, his library—everything except his Bible. He gave his money to the poor, leaving himself with so little that he often had to beg or borrow to obtain food. He wore an iron belt with points on the inside so that he was in constant discomfort and pushed the belt's spikes into his flesh whenever he found himself in danger of feeling happy. He denounced his studies of mathematics and science. Of his childhood fascination with geometry, he wrote, "I can scarcely remember that there is such a thing as geometry. I recognize geometry to be so useless . . . it is quite possible I shall never think of it again."[13]

Yet Pascal remained productive. In the years that followed the trance, he recorded his thoughts about God, religion, and life. Those thoughts were later published in a book titled *Pensées*, a work that is still in print today. And although Pascal had denounced mathematics, amid his vision of the futility of the worldly life is a mathematical exposition in which he trained his weapon of mathematical probability squarely on a question of theology and created a contribution just as important as his earlier work on the problem of points.

The mathematics in *Pensées* is contained in two manuscript pages covered on both sides by writing going in every direction and full of erasures and corrections. In those pages, Pascal detailed an analysis of the pros and cons of one's duty to God as if he were calculating mathematically the wisdom of a wager. His great innovation was his method of balancing those pros and cons, a concept that is today called mathematical expectation.

Pascal's argument went like this: Suppose you concede that you don't know whether or not God exists and therefore assign a 50 percent chance to either proposition. How should you weigh these odds when deciding whether to lead a pious life? If you act piously and God exists, Pascal argued, your gain—eternal happiness—is infinite. If, on the other hand, God does not exist, your loss, or negative return, is small—the sacrifices of piety. To weigh these possible gains and losses, Pascal proposed, you multiply the probability of each possible outcome by its payoff and add them all up, forming a kind of average or expected payoff. In other words, the mathematical expectation of your return on piety is one-half infinity (your gain if God exists) minus one-half a small number (your loss if he does not exist). Pascal knew enough about infinity to know that the answer to this calculation is infinite, and thus the expected return on piety is infinitely positive. Every reasonable person, Pascal concluded, should therefore follow the laws of God. Today this argument is known as Pascal's wager.

Expectation is an important concept not just in gambling but in all decision making. In fact, Pascal's wager is often considered the founding of the mathematical discipline of game theory, the quantitative study of optimal decision strategies in games. I must admit I find such thinking addictive, and so I sometimes carry it a bit too far. "How much does that parking meter cost?" I ask my son. The sign says 25¢. Yes, but 1 time in every 20 or so visits, I come back late and find a ticket, which runs $40, so the 25¢ cost of the meter is really just a cruel lure, I explain, because my real cost is $2.25. (The extra $2 comes from my 1 in 20 chance of getting a ticket multiplied by its $40 cost.) "How about our driveway," I ask my other son, "is it a toll road?" Well, we've lived at the house about 5 years, or roughly 2,400 times of backing down the driveway, and 3 times I've clipped my mirror on the protruding fence post at $400 a shot. You may as well put a toll box out there and toss in 50¢ each time you back up, he tells me. He understands expectation. (He also recommends that I refrain from driving them to school before I've had my morning coffee.)

Looking at the world through the lens of mathematical expectation, one often comes upon surprising results. For example, a recent sweepstakes sent through the mail offered a grand prize of $5 million.[14] All you had to do to win was mail in your entry. There was no limit on how many times you could enter, but each entry had to be mailed in separately. The sponsors were apparently expecting about 200 million entries, because the fine print said that the chances of winning were 1 in 200 million. Does it pay to enter this kind of "free sweepstakes offer"? Multiplying the probability of winning times the payoff, we find that each entry was worth $\frac{1}{40}$ of $1, or 2.5¢—far less than the cost of mailing it in. In fact, the big winner in this contest was the post office, which, if the projections were correct, made nearly $80 million in postage revenue on all the submissions.

Here's another crazy game. Suppose the state of California made its citizens the following offer: Of all those who pay the dollar or two to enter, most will receive nothing, one person will receive a fortune, and one person will be put to death in a violent manner. Would anyone enroll in that game? People do, and with enthusiasm. It is called the state lottery. And although the state does not advertise it in the manner in which I have described it, that is the way it works in practice. For while one lucky person wins the grand prize in each game, many millions of other contestants drive to and from their local ticket vendors to purchase their tickets, and some die in accidents along the way. Applying statistics from the National Highway Traffic Safety Administration and depending on such assumptions as how far each individual drives, how many tickets he or she buys, and how many people are involved in a typical accident, you find that a reasonable estimate of those fatalities is about one death per game.

State governments tend to ignore arguments about the possible bad effects of lotteries. That's because, for the most part, they know enough about mathematical expectation to arrange that for each ticket purchased, the expected winnings—the total prize money divided by the number of tickets sold—is less than the cost of the ticket. This generally leaves a tidy difference that can be diverted to state coffers. In 1992, however, some investors in Melbourne, Australia, noticed that the Virginia Lottery violated this principle.[15] The lottery involved picking 6 numbers from 1 to 44. Pascal's triangle, should we find one that goes that far, would show that there are 7,059,052 ways of choosing 6 numbers from a group of 44. The lottery jackpot was $27 million, and with second, third, and fourth prizes included, the pot grew to $27,918,561. The clever investors reasoned, if they bought one ticket with each of the possible 7,059,052 number combinations, the value of those tickets would equal the value of the pot. That made each ticket worth about $27.9 million divided by 7,059,052, or about $3.95. For what price was the state of Virginia, in all its wisdom, selling the tickets? The usual $1.

The Australian investors quickly found 2,500 small investors in Australia, New Zealand, Europe, and the United States willing to put up an average of $3,000 each. If the scheme worked, the yield on that investment would be about $10,800. There were some risks in their plan. For one, since they weren't the only ones buying tickets, it was possible that another player or even more than one other player would also choose the winning ticket, meaning they would have to split the pot. In the 170 times the lottery had been held, there was no winner 120 times, a single winner only 40 times, and two winners just 10 times. If those frequencies reflected accurately their odds, then the data suggested there was a 120 in 170 chance they would get the pot all to themselves, a 40 in 170 chance they would end up with half the pot, and a 10 in 170 chance they would win just a third of it. Recalculating their expected winnings employing Pascal's principle of mathematical expectation, they found them to be (120/170 × $27.9 million) + (40/170 × $13.95 million) + (10/170 × $6.975 million) = $23.4 million. That is $3.31 per ticket, a great return on a $1 expenditure even after expenses.

But there was another danger: the logistic nightmare of completing the purchase of all the tickets by the lottery deadline. That could lead to the expenditure of a significant portion of their funds with no significant prize to show for it.

The members of the investment group made careful preparations. They filled out 1.4 million slips by hand, as required by the rules, each slip good for five games. They placed groups of buyers at 125 retail outlets and obtained cooperation from grocery stores, which profited from each ticket they sold. The scheme got going just seventy-two hours before the deadline. Grocery-store employees worked in shifts to sell as many tickets as possible. One store sold 75,000 in the last forty-eight hours. A chain store accepted bank checks for 2.4 million tickets, assigned the work of printing the tickets among its stores, and hired couriers to gather them. Still, in the end, the group ran out of time: they had purchased just 5 million of the 7,059,052 tickets.

Several days passed after the winning ticket was announced, and no one came forward to present it. The consortium had won, but it took its members that long to find the winning ticket. Then, when state lottery officials discovered what the consortium had done, they balked at paying. A month of legal wrangling ensued before the officials concluded they had no valid reason to deny the group. Finally, they paid out the prize.

To the study of randomness, Pascal contributed both his ideas about counting and the concept of mathematical expectation. Who knows what else he might have discovered, despite his renouncing mathematics, if his health had held up. But it did not. In July 1662, Pascal became seriously ill. His physicians prescribed the usual remedies: they bled him and administered violent purges, enemas, and emetics. He improved for a while, and then the illness returned, along with severe headaches, dizziness, and convulsions. Pascal vowed that if he survived, he would devote his life to helping the poor and asked to be moved to a hospital for the incurable, in order that, if he died, he would be in their company. He did die, a few days later, in August 1662. He was thirty-nine. An autopsy found the cause of death to be a brain hemorrhage, but it also revealed lesions in his liver, stomach, and intestines that accounted for the illnesses that had plagued him throughout his life.

CHAPTER 5

The Dueling Laws of Large and Small Numbers

◇

I N THEIR WORK, Cardano, Galileo, and Pascal assumed that the probabilities relevant to the problems they tackled were known. Galileo, for example, assumed that a die has an equal chance of landing on any of its six faces. But how solid is such "knowledge"? The grand duke's dice were probably designed not to favor any face, but that doesn't mean fairness was actually achieved. Galileo could have tested his assumption by observing a number of tosses and recording how often each face came up. If he had repeated the test several times, however, he would probably have found a slightly different distribution each time, and even small deviations might have mattered, given the tiny differential he was asked to explain. In order to make the early work on randomness applicable to the real world, that issue had to be addressed: What is the connection between underlying probabilities and observed results? What does it mean, from a practical point of view, when we say the chances are 1 in 6 a die will land on 2? If it doesn't mean that in any series of tosses the die will land on the 2 exactly 1 time in 6, then on what do we base our belief that the chances of throwing a 2 really are 1 in 6? And what does it mean when a doctor says that a drug is 70 percent effective or has serious side effects in 1 percent of the cases or when a poll finds that a candidate has support of 36 percent of voters? These are deep questions, related to the very meaning of the concept of randomness, a concept mathematicians still like to debate.

I recently engaged in such a discussion one warm spring day with a statistician visiting from Hebrew University, Moshe, who sat across the lunch table from me at Caltech. Between spoonfuls of nonfat yogurt, Moshe espoused the opinion that truly random numbers do not exist. "There is no such thing," he said. "Oh, they publish charts and write computer programs, but they are just fooling themselves. No one has ever found a method of producing randomness that's any better than throwing a die, and throwing a die just won't do it."

Moshe waved his white plastic spoon at me. He was agitated now. I felt a connection between his feelings about randomness and his religious convictions. Moshe is an Orthodox Jew, and I know that many religious people have problems thinking God can allow randomness to exist. "Suppose you want a string of N random numbers between 1 and 6," he told me. "You throw a die N times and record the string of N numbers that comes up. Is that a random string?"

No, he claimed, because no one can make a perfect die. There will always be some faces that are favored and some that are disfavored. It might take 1,000 throws to notice the difference, or 1 billion, but eventually you will notice it. You'll see more 4s than 6s or maybe fewer. Any artificial device is bound to suffer from that flaw, he said, because human beings do not have access to perfection. That may be, but Nature does, and truly random events do occur on the atomic level. In fact, that is the very basis of quantum theory, and so we spent the rest of our lunch in a discussion of quantum optics.

Today cutting-edge quantum generators produce truly random numbers from the toss of Nature's perfect quantum dice. In the past the perfection necessary for randomness was indeed an elusive goal. One of the most creative approaches came from New York City's Harlem crime syndicates around 1920.[1] Needing a daily supply of five-digit random numbers for an illegal lottery, the racketeers thumbed their noses at the authorities by employing the last five digits of the U.S. Treasury balance. (At this writing the U.S. government is in debt by $8,995,800,515,946.50, or $29,679.02 per person, so today the racketeers could have obtained their five digits from the per capita debt!) Their so-called Treasury lottery ran afoul of not only criminal law, however, but also scientific law, for according to a rule called Benford's law, numbers arising in this cumulative fashion are not random but rather are biased in favor of the lower digits.

Benford's law was discovered not by a fellow named Benford but by the American astronomer Simon Newcomb. Around 1881, Newcomb noticed that the pages of books of logarithms that dealt with numbers beginning with the numeral 1 were dirtier and more frayed than the pages corresponding to numbers beginning with the numeral 2, and so on, down to the numeral 9, whose pages, in comparison, looked clean and new. Assuming that in the long run, wear was proportional to amount of use, Newcomb concluded from his observations that the scientists with whom he shared the book were working with data that reflected that distribution of digits. The law's current name arose after Frank Benford noticed the same thing, in 1938, when scrutinizing the log tables at the General Electric Research Laboratory in Schenectady, New York. But neither man proved the law. That didn't happen until 1995, in work by Ted Hill, a mathematician at the Georgia Institute of Technology.

According to Benford's law, rather than all nine digits' appearing with equal frequency, the number 1 should appear as the first digit in data about 30 percent of the time; the digit 2, about 18 percent of the time; and so on, down to the digit 9, which should appear as the first digit about 5 percent of the time. A similar law, though less pronounced, applies to later digits. Many types of data obey Benford's law, in particular, financial data. In fact, the law seems tailor-made for mining large amounts of financial data in search of fraud.

One famous application involved a young entrepreneur named Kevin Lawrence, who raised $91 million to create a chain of high-tech health clubs.[2] Engorged with cash, Lawrence raced into action, hiring a bevy of executives and spending his investors' money as quickly as he had raised it. That would have been fine except for one detail: he and his cohorts were spending most of the money not on the business but on personal items. And since several homes, twenty personal watercraft, forty-seven cars (including five Hummers, four Ferraris, three Dodge Vipers, two DeTomaso Panteras, and a Lamborghini Diablo), two Rolex watches, a twenty-one-carat diamond bracelet, a $200,000 samurai sword, and a commercial-grade cotton candy machine would have been difficult to explain as necessary business expenditures, Lawrence and his pals tried to cover their tracks by moving investors' money through a complex web of bank accounts and shell companies to give the appearance of a bustling and growing business. Unfortunately for them, a suspicious forensic accountant named Darrell Dorrell compiled a list of over 70,000 numbers representing their various checks and wire transfers and compared the distribution of digits with Benford's law. The numbers failed the test.[3] That, of course, was only the beginning of the investigation, but from there the saga unfolded predictably, ending the day before Thanksgiving 2003, when, flanked by his attorneys and clad in light blue prison garb, Kevin Lawrence was sentenced to twenty years without possibility of parole. The IRS has also studied Benford's law as a way to identify tax cheats. One researcher even applied the law to thirteen years of Bill Clinton's tax returns. They passed the test.[4]

Presumably neither the Harlem syndicate nor its customers noticed these regularities in their lottery numbers. But had people like Newcomb, Benford, or Hill played their lottery, in principle they could have used Benford's law to make favorable bets, earning a nice supplement to their scholar's salary.

In 1947, scientists at the Rand Corporation needed a large table of random digits for a more admirable purpose: to help find approximate solutions to certain mathematical equations employing a technique aptly named the Monte Carlo method. To generate the digits, they employed electronically generated noise, a kind of electronic roulette wheel. Is electronic noise random? That is a question as subtle as the definition of randomness itself.

In 1896 the American philosopher Charles Sanders Peirce wrote that a random sample is one "taken according to a precept or method which, being applied over and over again indefinitely, would in the long run result in the drawing of any one of a set of instances as often as any other set of the same number."[5] That is called the frequency interpretation of randomness. The main alternative to it is called the subjective interpretation. Whereas in the frequency interpretation you judge a sample by the way it turned out, in the subjective interpretation you judge a sample by the way it is produced. According to the subjective interpretation, a number or set of numbers is considered random if we either don't know or cannot predict how the process that produces it will turn out.

The difference between the two interpretations is more nuanced than it may seem. For example, in a perfect world a throw of a die would be random by the first definition but not by the second, since all faces would be equally probable but we could (in a perfect world) employ our exact knowledge of the physical conditions and the laws of physics to determine before each throw exactly how the die will land. In the imperfect real world, however, a throw of a die is random according to the second definition but not the first. That's because, as Moshe pointed out, owing to its imperfections, a die will not land on each face with equal frequency; nevertheless, because of our limitations we have no prior knowledge about any face being favored over any other.

In order to decide whether their table was random, the Rand scientists subjected it to various tests. Upon closer inspection, their system was shown to have biases, just like Moshe's archetypally imperfect dice.[6] The Rand scientists made some refinements to their system but never managed to completely banish the regularities. As Moshe said, complete chaos is ironically a kind of perfection. Still, the Rand numbers proved random enough to be useful, and the company published them in 1955 under the catchy title *A Million Random Digits*.

In their research the Rand scientists ran into a roulette-wheel problem that had been discovered, in some abstract way, almost a century earlier by an Englishman named Joseph Jagger.[7] Jagger was an engineer and a mechanic in a cotton factory in Yorkshire, and so he had an intuitive feel for the capabilities—and the shortcomings— of machinery and one day in 1873 turned his intuition and fertile mind from cotton to cash. How perfectly, he wondered, can the roulette wheels in Monte Carlo really work?

The roulette wheel—invented, at least according to legend, by Blaise Pascal as he was tinkering with an idea for a perpetual-motion machine—is basically a large bowl with partitions (called frets) that are shaped like thin slices of pie. When the wheel is spun, a marble first bounces along the rim of the bowl but eventually comes to rest in one of the compartments, which are numbered 1 through 36, plus 0 (and 00 on American roulette wheels). The bettor's job is simple: to guess in which compartment the marble will land. The existence of roulette wheels is pretty good evidence that legitimate psychics don't exist, for in Monte Carlo if you bet $1 on a compartment and the marble lands there, the house pays you $35 (plus your initial dollar). If psychics really existed, you'd see them in places like that, hooting and dancing and pushing wheelbarrows of cash down the street, and not on Web sites calling themselves Zelda Who Knows All and Sees All and offering twenty-four-hour free online love advice in competition with about 1.2 million other Web psychics (according to Google). For me both the future and, increasingly, the past unfortunately appear obscured by a thick fog. But I do know one thing: my chances of losing at European roulette are 36 out of 37; my chances of winning, 1 out of 37. That means that for every $1 I bet, the casino stands to win $(36/37 \times \$1) - (1/37 \times \$35)$. That comes to $1/37$ of a dollar, or about 2.7¢. Depending on my state of mind, it's either the price I pay for the enjoyment of watching a little marble bounce around a big shiny wheel or else the price I pay for the opportunity of having lightning strike me (in a good way). At least that is how it is supposed to work.

But does it? Only if the roulette wheels are perfectly balanced, thought Jagger, and he had worked with enough machines to share Moshe's point of view. He was willing to bet they weren't. So he gathered his savings, traveled to Monte Carlo, and hired six assistants, one for each of the casino's six roulette wheels. Every day his assistants observed the wheels, writing down every number that came up in the twelve hours the casino was open. Every night, back in his hotel room, Jagger analyzed the numbers. After six days, he had not detected any bias in five of the wheels, but on the sixth wheel nine numbers came up noticeably more often than the others. And so on the seventh day he headed to the casino and started to bet heavily on the nine favored numbers: 7, 8, 9, 17, 18, 19, 22, 28, and 29.

When the casino shut that night, Jagger was up $70,000. His winnings did not go without notice. Other patrons swarmed his table, tossing down their own cash to get in on a good thing. And casino inspectors were all over him, trying to decipher his system or, better, catch him cheating. By the fourth day of betting, Jagger had amassed $300,000, and the casino's managers were desperate to get rid of the mystery guy, or at least thwart his scheme. One imagines this being accomplished by a burly fellow from Brooklyn. Actually the casino employees did something far more clever.

On the fifth day, Jagger began to lose. His losing, like his winning, was not something you could spot immediately. Both before and after the casino's trick, he would win some and lose some, only now he lost more often than he won instead of the other way around. With the casino's small margin, it would take some pretty diligent betting to drain Jagger's funds, but after four days of sucking in casino money, he wasn't about to let up on the straw. By the time his change of luck deterred him, Jagger had lost half his fortune. One may imagine that by then his mood—not to mention the mood of his hangers-on—was sour. How could his scheme have suddenly failed?

Jagger at last made an astute observation. In the dozens of hours he had spent winning, he had come to notice a tiny scratch on the roulette wheel. This scratch was now absent. Had the casino kindly touched it up so that he could drive them to bankruptcy in style? Jagger guessed not and checked the other roulette wheels. One of them had a scratch. The casino managers had correctly guessed that Jagger's days of success were somehow related to the wheel he was playing, and so overnight they had switched wheels. Jagger relocated and again began to win. Soon he had pumped his winnings past where they had been, to almost half a million.

Unfortunately for Jagger, the casino's managers, finally zeroing in on his scheme, found a new way to thwart him. They decided to move the frets each night after closing, turning them along the wheel so that each day the wheel's imbalance would favor different numbers, numbers unknown to Jagger. Jagger started losing again and finally quit. His gambling career over, he left Monte Carlo with $325,000 in hand, about $5 million in today's dollars. Back home, he left his job at the mill and invested his money in real estate.

It may appear that Jagger's scheme had been a sure thing, but it wasn't. For even a perfectly balanced wheel will not come up on 0, 1, 2, 3, and so on, with exactly equal frequencies, as if the numbers in the lead would politely wait for the laggards to catch up. Instead, some numbers are bound to come up more often than average and others less often. And so even after six days of observations, there remained a chance that Jagger was wrong. The higher frequencies he observed for certain numbers may have arisen by chance and may not have reflected higher probabilities. That means that Jagger, too, had to face the question we raised at the start of this chapter: given a set of underlying probabilities, how closely can you expect your observations of a system to conform to those probabilities? Just as Pascal's work was done in the new climate of (the scientific) revolution, so this question would be answered in the midst of a revolution, this one in mathematics—the invention of calculus.

IN 1680 a great comet sailed through our neighborhood of the solar system, close enough that the tiny fraction of sunlight it reflected was sufficient to make it prominent in the night sky of our own planet. It was in that part of earth's orbit called November that the comet was first spotted, and for months afterward it remained an object of intense scrutiny, its path recorded in great detail. In 1687, Isaac Newton would use these data as an example of his inverse square law of gravity at work. And on one clear night in that parcel of land called Basel, Switzerland, another man destined for greatness was also paying attention. He was a young theologian who, gazing at the bright, hazy light of the comet, realized that it was mathematics, not the church, with which he wanted to occupy his life.[8] With that realization sprouted not just Jakob Bernoulli's own career change but also what would become the greatest family tree in the history of mathematics: in the century and a half between Jakob's birth and 1800 the Bernoulli family produced a great many offspring, about half of whom were gifted, including eight noted mathematicians, and three (Jakob, his younger brother Johann, and Johann's son Daniel) who are today counted as among the greatest mathematicians of all times.

Comets at the time were considered by theologians and the general public alike as a sign of divine anger, and God must have seemed pretty pissed off to create this one—it occupied more than half the visible sky. One preacher called it a "heavenly warning of the Allpowerful and Holy God written and placed before the powerless and unholy children of men." It portended, he wrote, "a noteworthy change in spirit or in worldly matters" for their country or town.[9] Jakob Bernoulli had another point of view. In 1681 he published a pamphlet titled *Newly Discovered Method of How the Path of a Comet or Tailed Star Can Be Reduced to Certain Fundamental Laws, and Its Appearance Predicted.*

Bernoulli had scooped Newton on the comet by six years. At least he would have scooped him had his theory been correct. It wasn't, but claiming publicly that comets follow natural law and not God's whim was a gutsy thing to do, especially given that the prior year— almost fifty years after Galileo's condemnation—the professor of mathematics at the University of Basel, Peter Megerlin, had been roundly attacked by theologians for accepting the Copernican system and had been banned from teaching it at the university. A forbidding schism lay between the mathematician-scientists and the theologians in Basel, and Bernoulli was parking himself squarely on the side of the scientists.

Bernoulli's talent soon brought the embrace of the mathematics community, and when Megerlin died, in late 1686, Bernoulli succeeded him as professor of mathematics. By then Bernoulli was working on problems connected with games of chance. One of his major influences was a Dutch mathematician and scientist, Christiaan

Huygens, who in addition to improving the telescope, being the first to understand Saturn's rings, creating the first pendulum clock (based on Galileo's ideas), and helping to develop the wave theory of light, had written a mathematical primer on probability inspired by the ideas of Pascal and Fermat.

For Bernoulli, Huygens's book was an inspiration. And yet he saw in the theory Huygens presented severe limitations. It might be sufficient for games of chance, but what about aspects of life that are more subjective? How can you assign a definite probability to the credibility of legal testimony? Or to who was the better golfer, Charles I of England or Mary, Queen of Scots? (Both were keen golfers.) Bernoulli believed that for rational decision making to be possible, there must be a reliable and mathematical way to determine probabilities. His view reflected the culture of the times, in which to conduct one's affairs in a manner that was consistent with probabilistic expectation was considered the mark of a reasonable person. But it was not just subjectivity that, in Bernoulli's opinion, limited the old theory of randomness. He also recognized that the theory was not designed for situations of ignorance, in which the probabilities of various outcomes could be defined in principle but in practice were not known. It is the issue I discussed with Moshe and that Jagger had to address: What are the odds that an imperfect die will come up with a 6? What are your chances of contracting the plague? What is the probability that your breastplate can withstand a thrust from your opponent's long sword? In both subjective and uncertain situations, Bernoulli believed it would be "insanity" to expect to have the sort of prior, or a priori, knowledge of probabilities envisioned in Huygens's book.[10]

Bernoulli saw the answer in the same terms that Jagger later would: instead of depending on probabilities being handed to us, we should discern them through observation. Being a mathematician, he sought to make the idea precise. Given that you view a certain number of roulette spins, how closely can you nail down the underlying probabilities, and with what level of confidence? We'll return to those questions in the next chapter, but they are not quite the questions Bernoulli was able to answer. Instead, he answered a closely related question: how well are underlying probabilities reflected in actual results? Bernoulli considered it obvious that we are justified in expecting that as we increase the number of trials, the observed frequencies will reflect—more and more accurately—their underlying probabilities. He certainly wasn't the first to believe that. But he was the first to give the issue a formal treatment, to turn the idea into a proof, and to quantify it, asking how many trials are necessary, and how sure can we be. He was also among the first to appreciate the importance of the new subject of calculus in addressing these issues.

THE YEAR Bernoulli was named professor in Basel proved to be a milestone year in the history of mathematics: it was the year in which Gottfried Leibniz published his revolutionary paper laying out the principles of integral calculus, the complement to his 1684 paper on differential calculus. Newton would publish his own version of the subject in 1687, in his *Philosophiae Naturalis Principia Mathematica*, or *Mathematical Principles of Natural Philosophy*, often referred to simply as *Principia*. These advances would hold the key to Bernoulli's work on randomness.

By the time they published, both Leibniz and Newton had worked on the subject for years, but their almost simultaneous publications begged for controversy over who should be credited for the idea. The great mathematician Karl Pearson (whom we shall encounter again in chapter 8) said that the reputation of mathematicians "stands for posterity largely not on what they did, but on what their contemporaries attributed to them."[11] Perhaps Newton and Leibniz would have agreed with that. In any case neither was above a good fight, and the one that ensued was famously bitter. At the time the outcome was mixed. The Germans and Swiss learned their calculus from Leibniz's work, and the English and many of the French from Newton's. From the modern standpoint there is very little difference between the two, but in the long run Newton's contribution is often emphasized because he appears to have truly had the idea earlier and because in *Principia* he employed his invention in the creation of modern physics, making *Principia* probably the greatest scientific book ever written. Leibniz, though, had developed a better notation, and it is his symbols that are often used in calculus today.

Neither man's publications were easy to follow. In addition to being the greatest book on science, Newton's *Principia* has also been called "one of the most inaccessible books ever written."[12] And Leibniz's work, according to one of Jakob Bernoulli's biographers, was "understood by no one"; it was not only unclear but also full of misprints. Jakob's brother Johann called it "an enigma rather than an explanation."[13] In fact, so incomprehensible were both works that scholars have speculated that both authors might have intentionally made their works difficult to understand to keep amateurs from dabbling. This enigmatic quality was an advantage for Jakob Bernoulli, though, for it did separate the wheat from the chaff, and his intellect fell into the former category. Hence once he had deciphered Leibniz's ideas, he possessed a weapon shared by only a handful of others in the entire world, and with it he could easily solve problems that were exceedingly difficult for others to attempt.

The set of concepts central to both calculus and Bernoulli's work is that of sequence, series, and limit. The term *sequence* means much the same thing to a mathematician as it does to anybody else: an ordered succession of elements, such as points or numbers. A series is simply the sum of a sequence of numbers. And loosely speaking, if the elements of a sequence seem to be heading somewhere—toward a particular endpoint or a particular number—then that is called the limit of the sequence.

Though calculus represents a new sophistication in the understanding of sequences, that idea, like so many others, had already been familiar to the Greeks. In the fifth century B.C., in fact, the Greek philosopher Zeno employed a curious sequence to formulate a paradox that is still debated among college philosophy students today, especially after a few beers. Zeno's paradox goes like this: Suppose a student wishes to step to the door, which is 1 meter away. (We choose a meter here for convenience, but the same argument holds for a mile or any other measure.) Before she arrives there, she first must arrive at the halfway point. But in order to reach the halfway point, she must first arrive halfway to the halfway point—that is, at the one-quarter-way point. And so on, ad infinitum. In other words, in order to reach her destination, she must travel this sequence of distances: ½ meter, ¼ meter, ⅛ meter, ¹⁄₁₆ meter, and so on. Zeno argued that because the sequence goes on forever, she has to traverse an *infinite* number of *finite* distances. That, Zeno said, must take an infinite amount of time. Zeno's conclusion: you can never get anywhere.

Over the centuries, philosophers from Aristotle to Kant have debated this quandary. Diogenes the Cynic took the empirical approach: he simply walked a few steps, then pointed out that things in fact do move. To those of us who aren't students of philosophy, that probably sounds like a pretty good answer. But it wouldn't have impressed Zeno. Zeno was aware of the clash between his logical proof and the evidence of his senses; it's just that, unlike Diogenes, what Zeno trusted was logic. And Zeno wasn't just spinning his wheels. Even Diogenes would have had to admit that his response leaves us facing a puzzling (and, it turns out, deep) question: if our sensory evidence is correct, then what is wrong with Zeno's logic?

Consider the sequence of distances in Zeno's paradox: ½ meter, ¼ meter, ⅛ meter, ¹⁄₁₆ meter, and so on (the increments growing ever smaller). This sequence has an infinite number of terms, so we cannot compute its sum by simply adding them all up. But we can notice that although the number of terms is infinite, those terms get successively smaller. Might there be a finite balance between the endless stream of terms and their endlessly diminishing size? That is precisely the kind of question we can address by employing the concepts of sequence, series, and limit. To see how it works, instead of trying to calculate how far the student went after the entire infinity of Zeno's intervals, let's take one interval at a time. Here are the student's distances after the first few intervals:

After the first interval: $\frac{1}{2}$ meter

After the second interval: $\frac{1}{2}$ meter + $\frac{1}{4}$ meter = $\frac{3}{4}$ meter

After the third interval: $\frac{1}{2}$ meter + $\frac{1}{4}$ meter + $\frac{1}{8}$ meter =
$\frac{7}{8}$ meter

After the fourth interval: $\frac{1}{2}$ meter + $\frac{1}{4}$ meter + $\frac{1}{8}$ meter +
$\frac{1}{16}$ meter = $\frac{15}{16}$ meter

There is a pattern in these numbers: $\frac{1}{2}$ meter, $\frac{3}{4}$ meter, $\frac{7}{8}$ meter, $\frac{15}{16}$ meter . . . The denominator is a power of two, and the numerator is one less than the denominator. We might guess from this pattern that after 10 intervals the student would have traveled $\frac{1,023}{1,024}$ meter; after 20 intervals, $\frac{1,048,575}{1,048,576}$ meter; and so on. The pattern makes it clear that Zeno is correct that the more intervals we include, the greater the sum of distances we obtain. But Zeno is not correct when he says that the sum is headed for infinity. Instead, the numbers seem to be approaching 1; or as a mathematician would say, 1 meter is the limit of this sequence of distances. That makes sense, because although Zeno chopped her trip into an infinite number of intervals, she had, after all, set out to travel just 1 meter.

Zeno's paradox concerns the amount of time it takes to make the journey, not the distance covered. If the student were forced to take individual steps to cover each of Zeno's intervals, she would indeed be in some time trouble (not to mention her having to overcome the difficulty of taking submillimeter steps)! But if she is allowed to move at constant speed without pausing at Zeno's imaginary checkpoints—and why not?—then the time it takes to travel each of Zeno's intervals is proportional to the distance covered in that interval, and so since the total distance is finite, as is the total time—and fortunately for all of us—motion is possible after all.

Though the modern concept of limits wasn't worked out until long after Zeno's life, and even Bernoulli's—it came in the nineteenth century[14]—it is this concept that informs the spirit of calculus, and it is in this spirit that Jakob Bernoulli attacked the relationship between probabilities and observation. In particular, Bernoulli investigated what happens in the limit of an arbitrarily large number of repeated observations. Toss a (balanced) coin 10 times and you might observe 7 heads, but toss it 1 zillion times and you'll most likely get very near 50 percent. In the 1940s a South African mathematician named John Kerrich decided to test this out in a practical experiment, tossing a coin what must have seemed like 1 zillion times—actually it was 10,000—and recording the results of each toss.[15] You might think Kerrich would have had better things to do, but he was a war prisoner at the time, having had the bad luck of being a visitor in Copenhagen when the Germans invaded Denmark in April 1940. According to Kerrich's data, after 100 throws he had only 44 percent heads, but by the time he reached 10,000, the number was much closer to half: 50.67 percent. How do you quantify this phenomenon? The answer to that question was Bernoulli's accomplishment.

According to the historian and philosopher of science Ian Hacking, Bernoulli's work "came before the public with a brilliant portent of all the things we know about it now; its mathematical profundity, its unbounded practical applications, its squirming duality and its constant invitation for philosophizing. Probability had fully emerged." In Bernoulli's more modest words, his study proved to be one of "novelty, as well as . . . high utility." It was also an effort, Bernoulli wrote, of "grave difficulty."[16] He worked on it for twenty years.

JAKOB BERNOULLI called the high point of his twenty-year effort his "golden theorem." Modern versions of it that differ in their technical nuance go by various names: Bernoulli's theorem, the law of large numbers, and the weak law of large numbers. The phrase *law of large numbers* is employed because, as we've said, Bernoulli's theorem concerns the way results reflect underlying probabilities when we make a large number of observations. But we'll stick with Bernoulli's terminology and call his theorem the golden theorem because we will be discussing it in its original form.[17]

Although Bernoulli's interest lay in real-world applications, some of his favorite examples involved an item not found in most households: an urn filled with colored pebbles. In one scenario, he envisioned the urn holding 3,000 white pebbles and 2,000 black ones, a ratio of 60 percent white to 40 percent black. In this example you conduct a series of blind drawings from the urn "with replacement"—that is, replacing each pebble before drawing the next in order not to alter the 3:2 ratio. The a priori chances of drawing a white pebble are then 3 out of 5, or 60 percent, and so in this example Bernoulli's central question becomes, how strictly should you expect the proportion of white pebbles drawn to hew to the 60 percent ratio, and with what probability?

The urn example is a good one because the same mathematics that describes drawing pebbles from an urn can be employed to describe any series of trials in which each trial has two possible outcomes, as long as those outcomes are random and the trials are independent of each other. Today such trials are called Bernoulli trials, and a series of Bernoulli trials is a Bernoulli process. When a random trial has two possible outcomes, one is often arbitrarily labeled "success" and the other "failure." The labeling is not meant to be literal and sometimes has nothing to do with the everyday meaning of the words—say, in the sense that if you can't wait to read on, this book is a success, and if you are using this book to keep yourself and your sweetheart warm after the logs burned down, it is a failure. Flipping a coin, deciding to vote for candidate A or candidate B, giving birth to a boy or girl, buying or not buying a product, being cured or not being cured, even dying or living are examples of Bernoulli trials. Actions that have multiple outcomes can also be modeled as Bernoulli trials if the question you are asking can be phrased in a way that has a yes or no answer, such as "Did the die land on the number 4?" or "Is there any ice left on the North Pole?" And so, although Bernoulli wrote about pebbles and urns, all his examples apply equally to these and many other analogous situations.

With that understanding we return to the urn, 60 percent of whose pebbles are white. If you draw 100 pebbles from the urn (with replacement), you might find that exactly 60 of them are white, but you might also draw just 50 white pebbles or 59. What are the chances that you will draw between 58 percent and 62 percent white pebbles? What are the chances you'll draw between 59 percent and 61 percent? How much more confident can you be if instead of 100, you draw 1,000 pebbles or 1 million? You can never be 100 percent certain, but can you draw enough pebbles to make the chances 99.9999 percent certain that you will draw, say, between 59.9 percent and 60.1 percent white pebbles? Bernoulli's golden theorem addresses questions such as these.

In order to apply the golden theorem, you must make two choices. First, you must specify your tolerance of error. How near to the underlying proportion of 60 percent are you demanding that your series of trials come? You must choose an interval, such as plus or minus 1 percent or 2 percent or 0.00001 percent. Second, you must specify your tolerance of uncertainty. You can never be 100 percent sure a trial will yield the result you are aiming for, but you can ensure that you will get a satisfactory result 99 times out of 100 or 999 out of 1,000.

The golden theorem tells you that it is always possible to draw enough pebbles to be almost certain that the percentage of white pebbles you draw will be near 60 percent no matter how demanding you want to be in your personal definition of *almost certain* and *near.* It also gives a numerical formula for calculating the number of trials that are "enough," given those definitions.

The first part of the law was a conceptual triumph, and it is the only part that survives in modern versions of the theorem. Concerning the second part—Bernoulli's formula—it is important to understand that although the golden theorem specifies a number of trials that is sufficient to meet your goals of confidence and accuracy, it does not say you can't accomplish those goals with fewer trials. That doesn't affect the first part of the theorem, for which it is enough to know simply that the number of trials specified is finite. But Bernoulli also intended the number given by his formula to be of practical use. Unfortunately, in most practical applications it isn't. For instance, here is a numerical example Bernoulli worked out himself, although I have changed the context: Suppose 60 percent of the voters in Basel support the mayor. How many people must you poll for the chances to be 99.9 percent that you will find the mayor's support to be between 58 percent and 62 percent—that is, for the result to be accurate within plus or minus 2 percent? (Assume, in order to be consistent with Bernoulli, that the people polled are chosen at random, but with replacement. In other words, it is possible that you poll a person more than once.) The answer is 25,550, which in Bernoulli's time was roughly the entire population of Basel. That this number was impractical wasn't lost on Bernoulli. He also knew that accomplished gamblers can intuitively guess their chances of success at a new game based on a sample of far fewer than thousands of trial games.

One reason Bernoulli's numerical estimate was so far from optimal was that his proof was based on many approximations. Another reason was that he chose 99.9 percent as his standard of certainty—that is, he required that he get the wrong answer (an answer that differed more than 2 percent from the true one) less than 1 time in 1,000. That is a very demanding standard. Bernoulli called it moral certainty, meaning the degree of certainty he thought a reasonable person would require in order to make a rational decision. It is perhaps a measure of how much the times have changed that today we've abandoned the notion of moral certainty in favor of the one we encountered in the last chapter, statistical significance, meaning that your answer will be wrong less than 1 time in 20.

With today's mathematical methods, statisticians have shown that in a poll like the one I described, you can achieve a statistically significant result with an accuracy of plus or minus 5 percent by polling only 370 subjects. And if you poll 1,000, you can achieve a 90 percent chance of coming within 2 percent of the true result (60 percent approval of Basel's mayor). But despite its limitations, Bernoulli's golden theorem was a milestone because it showed, at least in principle, that a large enough sample will almost certainly reflect the underlying makeup of the population being sampled.

. . .

The Dueling Laws of Large and Small Numbers

IN REAL LIFE we don't often get to observe anyone's or anything's performance over thousands of trials. And so if Bernoulli required an overly strict standard of certainty, in real-life situations we often make the opposite error: we assume that a sample or a series of trials is representative of the underlying situation when it is actually far too small to be reliable. For instance, if you polled exactly 5 residents of Basel in Bernoulli's day, a calculation like the ones we discussed in chapter 4 shows that the chances are only about 1 in 3 that you will find that 60 percent of the sample (3 people) supported the mayor.

Only 1 in 3? Shouldn't the true percentage of the mayor's supporters be the *most probable* outcome when you poll a sample of voters? In fact, 1 in 3 *is* the most probable outcome: the odds of finding 0, 1, 2, 4, or 5 supporters are lower than the odds of finding 3. Nevertheless, finding 3 supporters is not likely: because there are so many of those nonrepresentative possibilities, their combined odds add up to twice the odds that your poll accurately reflects the population. And so in a poll of 5 voters, 2 times out of 3 you will observe the "wrong" percentage. In fact, about 1 in 10 times you'll find that all the voters you polled agree on whether they like or dislike the mayor. And so if you paid any attention to a sample of 5, you'd probably severely over- or underestimate the mayor's true popularity.

The misconception—or the mistaken intuition—that a small sample accurately reflects underlying probabilities is so widespread that Kahneman and Tversky gave it a name: the law of small numbers.[18] The law of small numbers is not really a law. It is a sarcastic name describing the misguided attempt to apply the law of large numbers when the numbers aren't large.

If people applied the (untrue) law of small numbers only to urns, there wouldn't be much impact, but as we've said, many events in life are Bernoulli processes, and so our intuition often leads us to misinterpret what we observe. That is why, as I described in chapter 1, when people observe the handful of more successful or less successful years achieved by the Sherry Lansings and Mark Cantons of the world, they assume that their past performance accurately predicts their future performance.

Let's apply these ideas to an example I mentioned briefly in chapter 4: the situation in which two companies compete head-to-head or two employees within a company compete. Think now of the CEOs of the Fortune 500 companies. Let's assume that, based on their knowledge and abilities, each CEO has a certain probability of success each year (however his or her company may define that). And to make things simple, let's assume that for these CEOs successful years occur with the same frequency as the white pebbles or the mayor's supporters: 60 percent. (Whether the true number is a little higher or a little lower doesn't affect the thrust of this argument.) Does that mean we should expect, in a given five-year period, that a CEO will have precisely three good years?

No. As the earlier analysis showed, even if the CEOs all have a nice cut-and-dried 60 percent success rate, the chances that in a given five-year period a particular CEO's performance will reflect that underlying rate are only 1 in 3! Translated to the Fortune 500, that means that over the past five years about 333 of the CEOs would have exhibited performance that did not reflect their true ability. Moreover, we should expect, by chance alone, about 1 in 10 of the CEOs to have five winning or losing years in a row. What does this tell us? It is more reliable to judge people by analyzing their abilities than by glancing at the scoreboard. Or as Bernoulli put it, "One should not appraise human action on the basis of its results."[19]

Going against the law of small numbers requires character. For while anyone can sit back and point to the bottom line as justification, assessing instead a person's actual knowledge and actual ability takes confidence, thought, good judgment, and, well, guts. You can't just stand up in a meeting with your colleagues and yell, "Don't fire her. She was just on the wrong end of a Bernoulli series." Nor is it likely to win you friends if you stand up and say of the gloating fellow who just sold more Toyota Camrys than anyone else in the history of the dealership, "It was just a random fluctuation." And so it rarely happens. Executives' winning years are attributed to their brilliance, explained retroactively through incisive hindsight. And when people don't succeed, we often assume the failure accurately reflects the proportion with which their talents and their abilities fill the urn.

Another mistaken notion connected with the law of large numbers is the idea that an event is more or less likely to occur because it has or has not happened recently. The idea that the odds of an event with a fixed probability increase or decrease depending on recent occurrences of the event is called the gambler's fallacy. For example, if Kerrich landed, say, 44 heads in the first 100 tosses, the coin would not develop a bias toward tails in order to catch up! That's what is at the root of such ideas as "her luck has run out" and "He is due." That does not happen. For what it's worth, a good streak doesn't jinx you, and a bad one, unfortunately, does not mean better luck is in store. Still, the gambler's fallacy affects more people than you might think, if not on a conscious level then on an unconscious one. People expect good luck to follow bad luck, or they worry that bad will follow good.

I remember, on a cruise a few years back, watching an intense pudgy man sweating as he frantically fed dollars into a slot machine as fast as it would take them. His companion, seeing me eye them, remarked simply, "He is due." Although tempted to point out that, *no, he isn't due,* I instead walked on. After several steps I halted my progress owing to a sudden flashing of lights, ringing of bells, not a little hooting on the couple's part, and the sound of, for what seemed like minutes, a fast stream of dollar coins flying out of the machine's chute. Now I know that a modern slot machine is computerized, its payoffs driven by a random-number generator, which by both law and regulation must truly generate, as advertised, random numbers, making each pull of the handle completely independent of the history of previous pulls. And yet . . . Well, let's just say the gambler's fallacy is a powerful illusion.

THE MANUSCRIPT in which Bernoulli presented his golden theorem ends abruptly even though he promises earlier in the work that he will provide applications to various issues in civic affairs and economics. It is as if "Bernoulli literally quit when he saw the number 25,550," wrote the historian of statistics Stephen Stigler.[20] In fact, Bernoulli was in the process of publishing his manuscript when he died "of a slow fever" in August 1705, at the age of fifty. His publishers asked Johann Bernoulli to complete it, but Johann refused, saying he was too busy. That may appear odd, but the Bernoullis were an odd family. If you were asked to choose the most unpleasant mathematician who ever lived, you wouldn't be too far off if you fingered Johann Bernoulli. He has been variously described in historical texts as jealous, vain, thin-skinned, stubborn, bilious, boastful, dishonest, and a consummate liar. He accomplished much in mathematics, but he is also known for having his son Daniel tossed out of the Académie des Sciences after Daniel won a prize for which Johann himself had competed, for attempting to steal both his brother's and Leibniz's ideas, and for plagiarizing Daniel's book on hydrodynamics and then faking the publication date so that his book would appear to have been published first.

When he was asked to complete his late brother's manuscript, he had recently relocated to Basel from the University of Groningen, in the Netherlands, obtaining a post not in mathematics but as a professor of Greek. Jakob had found this career change suspicious, especially since in his estimation Johann did not know Greek. What Jakob suspected, he wrote Leibniz, was that Johann had come to Basel to usurp Jakob's position. And, indeed, upon Jakob's death, Johann did obtain it.

Johann and Jakob had not gotten along for most of their adult lives. They would regularly trade insults in mathematics publications and in letters that, one mathematician wrote, "bristle with strong language that is usually reserved for horse thieves."[21] And so when the need arose to edit Jakob's posthumous manuscript, the task fell further down the food chain, to Jakob's nephew Nikolaus, the son of one of Jakob's other brothers, also named Nikolaus. The younger Nikolaus was only eighteen at the time, but he had been one of Jakob's pupils. Unfortunately he didn't feel up to the task, possibly in part because he was aware of Leibniz's opposition to his uncle's ideas about applications of the theory. And so the manuscript lay dormant for eight years. The book was finally published in 1713 under the title *Ars conjectandi*, or *The Art of Conjecture*. Like Pascal's *Pensées*, it is still in print.

Jakob Bernoulli had shown that through mathematical analysis one could learn how the inner hidden probabilities that underlie natural systems are reflected in the data those systems produce. As for the question that Bernoulli did not answer—the question of how to infer, from the data produced, the underlying probability of events—the answer would not come for several decades more.

CHAPTER 6

False Positives and Positive Fallacies

I N THE 1970s a psychology professor at Harvard had an odd-
looking middle-aged student in his class. After the first few class
meetings the student approached the professor to explain why he
had enrolled in the class.[1] In my experience teaching, though I have
had some polite students come up to me to explain why they were
dropping my course, I have never had a student feel the need to
explain why he was taking it. That's probably why I can get away with
happily assuming that if asked, such a student would respond,
"Because I am fascinated by the subject, and you are a fine lecturer."
But this student had other reasons. He said he needed help because
strange things were happening to him: his wife spoke the words he
was thinking before he could say them, and now she was divorcing
him; a co-worker casually mentioned layoffs over drinks, and two
days later the student lost his job. Over time, he reported, he had
experienced dozens of misfortunes and what he considered to be dis-
turbing coincidences.

At first the happenings confused him. Then, as most of us would,
he formed a mental model to reconcile the events with the way he
believed the world behaves. The theory he came up with, however,
was unlike anything most of us would devise: he was the subject of an
elaborate secret scientific experiment. He believed the experiment

was staged by a large group of conspirators led by the famous psychologist B. F. Skinner. He also believed that when it was over, he would become famous and perhaps be elected to a high public office. That, he said, was why he was taking the course. He wanted to learn how to test his hypothesis in light of the many instances of evidence he had accumulated.

A few months after the course had run its course, the student again called on the professor. The experiment was still in progress, he reported, and now he was suing his former employer, who had produced a psychiatrist willing to testify that he suffered from paranoia.

One of the paranoid delusions the former employer's psychiatrist pointed to was the student's alleged invention of a fictitious eighteenth-century minister. In particular, the psychiatrist scoffed at the student's claim that this minister was an amateur mathematician who had created in his spare moments a bizarre theory of probability. The minister's name, according to the student, was Thomas Bayes. His theory, the student asserted, described how to assess the chances that some event would occur if some other event also occurred. What are the chances that a particular student would be the subject of a vast secret conspiracy of experimental psychologists? Admittedly not huge. But what if one's wife speaks one's thoughts before one can utter them *and* co-workers foretell your professional fate over drinks in casual conversation? The student claimed that Bayes's theory showed how you should alter your initial estimation in light of that new evidence. And he presented the court with a mumbo jumbo of formulas and calculations regarding his hypothesis, concluding that the additional evidence meant that the probability was 999,999 in 1 million that he was right about the conspiracy. The enemy psychiatrist claimed that this mathematician-minister and his theory were figments of the student's schizophrenic imagination.

The student asked the professor to help him refute that claim. The professor agreed. He had good reason, for Thomas Bayes, born in London in 1701, really was a minister, with a parish at Tunbridge Wells. He died in 1761 and was buried in a park in London called Bunhill Fields, in the same grave as his father, Joshua, also a minister. And he indeed did invent a theory of "conditional probability" to show how the theory of probability can be extended from independent events to events whose outcomes are connected. For example, the probability that a randomly chosen person is mentally ill and the probability that a randomly chosen person believes his spouse can read his mind are both very low, but the probability that a person is mentally ill *if* he believes his spouse can read his mind is much higher, as is the probability that a person believes his spouse can read his mind *if* he is mentally ill. How are all these probabilities related? That question is the subject of conditional probability.

The professor supplied a deposition explaining Bayes's existence and his theory, though not supporting the specific and dubious calculations that his former student claimed proved his sanity. The sad part of this story is not just the middle-aged schizophrenic himself, but the medical and legal team on the other side. It is unfortunate that some people suffer from schizophrenia, but even though drugs can help to mediate the illness, they cannot battle ignorance. And ignorance of the ideas of Thomas Bayes, as we shall see, resides at the heart of many serious mistakes in both medical diagnosis and legal judgment. It is an ignorance that is rarely addressed during a doctor's or a lawyer's professional training.

We also make Bayesian judgments in our daily lives. A film tells the story of an attorney who has a great job, a charming wife, and a wonderful family. He loves his wife and daughter, but still he feels that something is missing in his life. One night as he returns home on the train he spots a beautiful woman gazing with a pensive expression out the window of a dance studio. He looks for her again the next night, and the night after that. Each night as his train passes her studio, he falls further under her spell. Finally one evening he impulsively rushes off the train and signs up for dance lessons, hoping to meet the woman. He finds that her haunting attraction withers once his gaze from afar gives way to face-to-face encounters. He does fall in love, however, not with her but with dancing.

He keeps his new obsession from his family and colleagues, making excuses for spending more and more evenings away from home.

His wife eventually discovers that he is not working late as often as he says he is. She figures the chances of his lying about his after-work activities are far greater if he is having an affair than if he isn't, and so she concludes that he is. But the wife was mistaken not just in her conclusion but in her reasoning: she confused the probability that her husband would sneak around *if* he were having an affair with the probability that he was having an affair *if* he was sneaking around.

It's a common mistake. Say your boss has been taking longer than usual to respond to your e-mails. Many people would take that as a sign that their star is falling because *if* your star is falling, the chances are high that your boss will respond to your e-mails more slowly than before. But your boss might be slower in responding because she is unusually busy or her mother is ill. And so the chances that your star is falling *if* she is taking longer to respond are much lower than the chances that your boss will respond more slowly *if* your star is falling. The appeal of many conspiracy theories depends on the misunderstanding of this logic. That is, it depends on confusing the probability that a series of events would happen *if* it were the product of a huge conspiracy with the probability that a huge conspiracy exists *if* a series of events occurs.

The effect on the probability that an event will occur *if* or *given that* other events occur is what Bayes's theory is all about. To see in detail how it works, we'll turn to another problem, one that is related to the two-daughter problem we encountered in chapter 3. Let us now suppose that a distant cousin has two children. Recall that in the two-daughter problem you know that one or both are girls, and you are trying to remember which it is—one or both? In a family with two children, what are the chances, if one of the children is a girl, that both children are girls? We didn't discuss the question in those terms in chapter 3, but the *if* makes this a problem in conditional probability. If that *if* clause were not present, the chances that both children were girls would be 1 in 4, the 4 possible birth orders being (boy, boy), (boy, girl), (girl, boy), and (girl, girl). But given the additional information that the family has a girl, the chances are 1 in 3. That is because if one of the children is a girl, there are just 3 possible scenarios for this family—(boy, girl), (girl, boy), and (girl, girl)—and exactly 1 of the 3 corresponds to the outcome that both children are girls. That's probably the simplest way to look at Bayes's ideas—they are just a matter of accounting. First write down the sample space—that is, the list of all the possibilities—along with their probabilities if they are not all equal (that is actually a good idea in analyzing any confusing probability issue). Next, cross off the possibilities that the condition (in this case, "at least one girl") eliminates. What is left are the remaining possibilities and their relative probabilities.

That might all seem obvious. Feeling cocky, you may think you could have figured it out without the help of dear Reverend Bayes and vow to grab a different book to read the next time you step into the bathtub. So before we proceed, let's try a slight variant on the two-daughter problem, one whose resolution may be a bit more shocking.[2]

The variant is this: in a family with two children, what are the chances, if one of the children is a girl named Florida, that both children are girls? Yes, I said a girl named Florida. The name might sound random, but it is not, for in addition to being the name of a state known for Cuban immigrants, oranges, and old people who traded their large homes up north for the joys of palm trees and organized bingo, it is a real name. In fact, it was in the top 1,000 female American names for the first thirty or so years of the last century. I picked it rather carefully, because part of the riddle is the question, what, if anything, about the name Florida affects the odds? But I am getting ahead of myself. Before we move on, please consider this question: in the girl-named-Florida problem, are the chances of two girls still 1 in 3 (as they are in the two-daughter problem)?

I will shortly show that the answer is no. The fact that one of the girls is named Florida changes the chances to 1 in 2: Don't worry if that is difficult to imagine. The key to understanding randomness and all of mathematics is not being able to intuit the answer to every problem immediately but merely having the tools to figure out the answer.

THOSE WHO DOUBTED Bayes's existence were right about one thing: he never published a single scientific paper. We know little of his life, but he probably pursued his work for his own pleasure and did not feel much need to communicate it. In that and other respects he and Jakob Bernoulli were opposites. For Bernoulli resisted the study of theology, whereas Bayes embraced it. And Bernoulli sought fame, whereas Bayes showed no interest in it. Finally, Bernoulli's theorem concerns how many heads to expect if, say, you plan to conduct many tosses of a balanced coin, whereas Bayes investigated Bernoulli's original goal, the issue of how certain you can be that a coin is balanced if you observe a certain number of heads.

The theory for which Bayes is known today came to light on December 23, 1763, when another chaplain and mathematician, Richard Price, read a paper to the Royal Society, Britain's national academy of science. The paper, by Bayes, was titled "An Essay toward Solving a Problem in the Doctrine of Chances" and was published in the Royal Society's *Philosophical Transactions* in 1764. Bayes had left Price the article in his will, along with £100. Referring to Price as "I suppose a preacher at Newington Green," Bayes died four months after writing his will.[3]

Despite Bayes's casual reference, Richard Price was not just another obscure preacher. He was a well-known advocate of freedom of religion, a friend of Benjamin Franklin's, a man entrusted by Adam Smith to critique parts of a draft of *The Wealth of Nations*, and a well-known mathematician. He is also credited with founding actuary science, a field he developed when, in 1765, three men from an insurance company, the Equitable Society, requested his assistance. Six years after that encounter he published his work in a book titled *Observations on Reversionary Payments*. Though the book served as a bible for actuaries well into the nineteenth century, because of some poor data and estimation methods, he appears to have underestimated life expectancies. The resulting inflated life insurance premiums enriched his pals at the Equitable Society. The hapless British government, on the other hand, based annuity payments on Price's tables and took a bath when the pensioners did not proceed to keel over at the predicted rate.

As I mentioned, Bayes developed conditional probability in an attempt to answer the same question that inspired Bernoulli: how can we infer underlying probability from observation? *If a drug just cured 45 out of 60 patients in a clinical trial, what does that tell you about the chances the drug will work on the next patient? If it worked for 600,000 out of 1 million patients, the odds are obviously good that its chances of working are close to 60 percent. But what can you conclude from a smaller trial? Bayes also asked another question: if, before the trial, you had reason to believe that the drug was only 50 percent effective, how much weight should the new data carry in your future assessments? Most of our life experiences are like that: we observe a relatively small sample of outcomes, from which we infer information and make judgments about the qualities that produced those outcomes. How should we make those inferences?

Bayes approached the problem via a metaphor.[4] Imagine we are supplied with a square table and two balls. We roll the first ball onto the table in a manner that makes it equally probable that the ball will come to rest at any point. Our job is to determine, without looking, where along the left-right axis the ball stopped. Our tool in this is the second ball, which we may repeatedly roll onto the table in the same manner as the first. With each roll a collaborator notes whether that ball comes to rest to the right or the left of the place where the first ball landed. At the end he informs us of the total number of times the second ball landed in each of the two general locations. The first ball represents the unknown that we wish to gain information about, and the second ball represents the evidence we manage to obtain. If the second ball lands consistently to the right of the first, we can be pretty confident that the first ball rests toward the far left side of the table. If it lands less consistently to the right, we might be less confident of that conclusion, or we might guess that the first ball is situated farther to the right. Bayes showed how to determine, based on the data of the second ball, the precise probability that the first ball is at any given point on the left-right axis. And he showed how, given additional data, one should revise one's initial estimate. In Bayesian terminology the initial estimates are called prior probabilities and the new guesses, posterior probabilities.

Bayes concocted this game because it models many of the decisions we make in life. In the drug-trial example the position of the first ball represents the drug's true effectiveness, and the reports regarding the second ball represent the patient data. The position of the first ball could also represent a film's appeal, product quality, driving skill, hard work, stubbornness, talent, ability, or whatever it is that determines the success or failure of a certain endeavor. The reports on the second ball would then represent our observations or the data we collect. Bayes's theory shows how to make assessments and then adjust them in the face of new data.

Today Bayesian analysis is widely employed throughout science and industry. For instance, models employed to determine car insurance rates include a mathematical function describing, per unit of driving time, your personal probability of having zero, one, or more accidents. Consider, for our purposes, a simplified model that places everyone in one of two categories: high risk, which includes drivers who average at least one accident each year, and low risk, which includes drivers who average less than one. If, when you apply for insurance, you have a driving record that stretches back twenty years without an accident or one that goes back twenty years with thirty-seven accidents, the insurance company can be pretty sure which category to place you in. But if you are a new driver, should you be classified as low risk (a kid who obeys the speed limit and volunteers to be the designated driver) or high risk (a kid who races down Main Street swigging from a half-empty $2 bottle of Boone's Farm apple wine)? Since the company has no data on you—no idea of the "position of the first ball"—it might assign you an equal prior probability of being in either group, or it might use what it knows about the general population of new drivers and start you off by guessing that the chances you are a high risk are, say, 1 in 3. In that case the company would model you as a hybrid—one-third high risk and two-thirds low risk—and charge you one-third the price it charges high-risk drivers plus two-thirds the price it charges low-risk drivers. Then, after a year of observation—that is, after one of Bayes's second balls has been thrown—the company can employ the new datum to reevaluate its model, adjust the one-third and two-third proportions it previously assigned, and recalculate what it ought to charge. If you have had no accidents, the proportion of low risk and low price it assigns you will increase; if you have had two accidents, it will decrease. The precise size of the adjustment is given by Bayes's theory. In the same manner the insurance company can periodically adjust its assessments in later years to reflect the fact that you were accident-free or that you twice had an accident while driving the wrong way down a one-way street, holding a cell phone with your left hand and a doughnut with your right. That is why insurance companies can give out "good driver" discounts: the absence of accidents elevates the posterior probability that a driver belongs in a low-risk group.

Obviously many of the details of Bayes's theory are rather complex. But as I mentioned when I analyzed the two-daughter problem, the key to his approach is to use new information to prune the sample space and adjust probabilities accordingly. In the two-daughter problem the sample space was initially (boy, boy), (boy, girl), (girl, boy), and (girl, girl) but reduces to (boy, girl), (girl, boy), and (girl, girl) *if* you learn that one of the children is a girl, making the chances of a two-girl family 1 in 3. Let's apply that same simple strategy to see what happens if you learn that one of the children is a girl named Florida.

In the girl-named-Florida problem our information concerns not just the gender of the children, but also, for the girls, the name. Since our original sample space should be a list of all the possibilities, in this case it is a list of both gender and name. Denoting "girl-named-Florida" by girl-F and "girl-not-named-Florida" by girl-NF, we write the sample space this way: (boy, boy), (boy, girl-F), (boy, girl-NF), (girl-F, boy), (girl-NF, boy), (girl-NF, girl-F), (girl-F, girl-NF), (girl-NF, girl-NF), and (girl-F, girl-F).

Now, the pruning. Since we know that one of the children is a girl named Florida, we can reduce the sample space to (boy, girl-F), (girl-F, boy), (girl-NF, girl-F), (girl-F, girl-NF), and (girl-F, girl-F). That brings us to another way in which this problem differs from the two-daughter problem. Here, because it is not equally probable that a girl's name is or is not Florida, not all the elements of the sample space are equally probable.

In 1935, the last year for which the Social Security Administration provided statistics on the name, about 1 in 30,000 girls were christened Florida.[5] Since the name has been dying out, for the sake of argument let's say that today the probability of a girl's being named Florida is 1 in 1 million. That means that if we learn that a particular girl's name is not Florida, it's no big deal, but if we learn that a particular girl's name is Florida, in a sense we've hit the jackpot. The chances of both girls' being named Florida (even if we ignore the fact that parents tend to shy away from giving their children identical names) are therefore so small we are justified in ignoring that possibility. That leaves us with just (boy, girl-F), (girl-F, boy), (girl-NF, girl-F), and (girl-F, girl-NF), which are, to a very good approximation, equally likely.

Since 2 of the 4, or half, of the elements in the sample space are families with two girls, the answer is not 1 in 3—as it was in the two-daughter problem—but 1 in 2. The added information—your knowledge of the girl's name—makes a difference.

One way to understand this, if it still seems puzzling, is to imagine that we gather into a very large room 75 million families that have two children, at least one of whom is a girl. As the two-daughter problem taught us, there will be about 25 million two-girl families in that room and 50 million one-girl families (25 million in which the girl is the older child and an equal number in which she is the younger). Next comes the pruning: we ask that only the families that include a girl named Florida remain. Since Florida is a 1 in 1 million name, about 50 of the 50 million one-girl families will remain. And of the 25 million two-girl families, 50 of them will also get to stay, 25 because their firstborn is named Florida and another 25 because their younger girl has that name. It's as if the girls are lottery tickets and the girls named Florida are the winning tickets. Although there are twice as many one-girl families as two-girl families, the two-girl families each have two tickets, so the one-girl families and the two-girl families will be about equally represented among the winners.

I have described the girl-named-Florida problem in potentially annoying detail, the kind of detail that sometimes lands me on the do-not-invite list for my neighbors' parties. I did this not because I expect you to run into this situation. I did it because the context is simple, and the same kind of reasoning will bring clarity to many situations that really are encountered in life. Now let's talk about a few of those.

My most memorable encounter with the Reverend Bayes came one Friday afternoon in 1989, when my doctor told me by telephone that the chances were 999 out of 1,000 that I'd be dead within a decade. He added, "I'm *really* sorry," as if he had some patients to whom he would say he is sorry but not mean it. Then he answered a few questions about the course of the disease and hung up, presumably to offer another patient his or her Friday-afternoon news flash. It is hard to describe or even remember exactly how the weekend went for me, but let's just say I did not go to Disneyland. Given my death sentence, why am I still here, able to write about it?

The adventure started when my wife and I applied for life insurance. The application procedure involved a blood test. A week or two later we were turned down. The ever economical insurance company sent the news in two brief letters that were identical, except for a single additional word in the letter to my wife. My letter stated that the company was denying me insurance because of the "results of your blood test." My wife's letter stated that the company was turning her down because of the "results of your husband's blood test." When

114

the added word *husband's* proved to be the extent of the clues the kindhearted insurance company was willing to provide about our uninsurability, I went to my doctor on a hunch and took an HIV test. It came back positive. Though I was too shocked initially to quiz him about the odds he quoted, I later learned that he had derived my 1-in-1,000 chance of being healthy from the following statistic: the HIV test produced a positive result when the blood was not infected with the AIDS virus in only 1 in 1,000 blood samples. That might sound like the same message he passed on, but it wasn't. My doctor had confused the chances that I would test positive *if* I was not HIV-positive with the chances that I would not be HIV-positive *if* I tested positive.

To understand my doctor's error, let's employ Bayes's method. The first step is to define the sample space. We could include everyone who has ever taken an HIV test, but we'll get a more accurate result if we employ a bit of additional relevant information about me and consider only heterosexual non-IV-drug-abusing white male Americans who have taken the test. (We'll see later what kind of difference this makes.)

Now that we know whom to include in the sample space, let's classify the members of the space. Instead of boy and girl, here the relevant classes are those who tested positive and are HIV-positive (true positives), those who tested positive but are not positive (false positives), those who tested negative and are HIV-negative (true negatives), and those who tested negative but are HIV-positive (false negatives).

Finally, we ask, how many people are there in each of these classes? Suppose we consider an initial population of 10,000. We can estimate, employing statistics from the Centers for Disease Control and Prevention, that in 1989 about 1 in those 10,000 heterosexual non-IV-drug-abusing white male Americans who got tested were infected with HIV.[6] Assuming that the false-negative rate is near 0, that means that about 1 person out of every 10,000 will test positive due to the presence of the infection. In addition, since the rate of false positives is, as my doctor had quoted, 1 in 1,000, there will be about 10 others who are not infected with HIV but will test positive anyway. The other 9,989 of the 10,000 men in the sample space will test negative.

Now let's prune the sample space to include only those who tested positive. We end up with 10 people who are false positives and 1 true positive. In other words, only 1 in 11 people who test positive are really infected with HIV. My doctor told me that the probability that the test was wrong—and I was in fact healthy—was 1 in 1,000. He should have said, "Don't worry, the chances are better than 10 out of 11 that you are not infected." In my case the screening test was apparently fooled by certain markers that were present in my blood even though the virus this test was screening for was not present.

It is important to know the false positive rate when assessing any diagnostic test. For example, a test that identifies 99 percent of all malignant tumors sounds very impressive, but I can easily devise a test that identifies 100 percent of all tumors. All I have to do is report that everyone I examine has a tumor. The key statistic that differentiates my test from a useful one is that my test would produce a high rate of false positives. But the above incident illustrates that knowledge of the false positive rate is not sufficient to determine the usefulness of a test—you must also know how the false-positive rate compares with the true prevalence of the disease. If the disease is rare, even a low false-positive rate does not mean that a positive test implies you have the disease. If a disease is common, a positive result is much more likely to be meaningful. To see how the true prevalence affects the implications of a positive test, let's suppose now that I had been homosexual and tested positive. Assume that in the male gay community the chance of infection among those being tested in 1989 was about 1 percent. That means that in the results of 10,000 tests, we would find not 1 (as before), but 100 true positives to go with the 10 false positives. So in this case the chances that a positive test meant I was infected would have been 10 out of 11. That's why, when assessing test results, it is good to know whether you are in a high-risk group.

BAYES'S THEORY shows that the probability that A will occur if B occurs will generally differ from the probability that B will occur if A occurs.[7] To not account for this is a common mistake in the medical profession. For instance, in studies in Germany and the United States, researchers asked physicians to estimate the probability that an asymptomatic woman between the ages of 40 and 50 who has a positive mammogram actually has breast cancer if 7 percent of mammograms show cancer when there is none.[8] In addition, the doctors were told that the actual incidence was about 0.8 percent and that the false-negative rate about 10 percent. Putting that all together, one can use Bayes's methods to determine that a positive mammogram is due to cancer in only about 9 percent of the cases. In the German group, however, one-third of the physicians concluded that the probability was about 90 percent, and the median estimate was 70 percent. In the American group, 95 out of 100 physicians estimated the probability to be around 75 percent.

Similar issues arise in drug testing in athletes. Here again, the oft-quoted but not directly relevant number is the false positive rate. This gives a distorted view of the probability that an athlete is guilty. For example, Mary Decker Slaney, a world-class runner and 1983 world champion in the 1,500 and 3,000 meter race, was trying to make a comeback when, at the U.S. Olympic Trials in Atlanta in 1996, she was accused of doping violations consistent with testosterone use. After various deliberations, the IAAF (known officially since 2001 as the International Association of Athletics Federations) ruled that Slaney "was guilty of a doping offense," effectively ending her career. According to some of the testimony in the Slaney case the false-positive rate for the test to which her urine was subjected could have been as high as 1 percent. This probably made many people comfortable that her chance of guilt was 99 percent, but as we have seen that is not true. Suppose, for example, 1,000 athletes were tested, 1 in 10 was guilty, and the test, when given to a guilty athlete, had a 50 percent chance of revealing the doping violation. Then for every thousand athletes tested, 100 would have been guilty and the test would have fingered 50 of those. Meanwhile, of the 900 athletes who are innocent, the test would have fingered 9. So what a positive-doping test really meant was not that the probability she was guilty was 99 percent, but rather $^{50}/_{59}$ = 84.7 percent. Put another way, you should have about as much confidence that Slaney was guilty based on that evidence as you would that the number 1 won't turn up when she tossed a die. That certainly leaves room for reasonable doubt, and, more important, indicates that to perform mass testing (90,000 athletes have their urine tested annually) and make judgments based on such a procedure means to condemn a large number of innocent people.[9]

In legal circles the mistake of inversion is sometimes called the prosecutor's fallacy because prosecutors often employ that type of fallacious argument to lead juries to convicting suspects on thin evidence. Consider, for example, the case in Britain of Sally Clark.[10] Clark's first child died at 11 weeks. The death was reported as due to sudden infant death syndrome, or SIDS, a diagnosis that is made when the death of a baby is unexpected and a postmortem does not reveal a cause of death. Clark conceived again, and this time her baby died at 8 weeks, again reportedly of SIDS. When that happened, she was arrested and accused of smothering both children. At the trial the prosecution called in an expert pediatrician, Sir Roy Meadow, to testify that based on the rarity of SIDS, the odds of both children's dying from it was 73 million to 1. The prosecution offered no other substantive evidence against her. Should that have been enough to convict? The jury thought so, and in November 1999, Mrs. Clark was sent to prison.

Sir Meadow had estimated that the odds that a child will die of SIDS are 1 in 8,543. He calculated his estimate of 73 million to 1 by multiplying two such factors, one for each child. But this calculation assumes that the deaths are independent—that is, that no environmental or genetic effects play a role that might increase a second child's risk once an older sibling has died of SIDS. In fact, in an editorial in the *British Medical Journal* a few weeks after the trial, the chances of two siblings' dying of SIDS were estimated at 2.75 million to 1.[11] Those are still very long odds.

The key to understanding why Sally Clark was wrongly imprisoned is again to consider the inversion error: it is not the probability that two children will die of SIDS that we seek but the probability that the two children who died, died of SIDS. Two years after Clark was imprisoned, the Royal Statistical Society weighed in on this subject with a press release, declaring that the jury's decision was based on "a serious error of logic known as the Prosecutor's Fallacy. The jury needs to weigh up two competing explanations for the babies' deaths: SIDS or murder. Two deaths by SIDS or two murders are each quite unlikely, but one has apparently happened in this case. What matters is the relative likelihood of the deaths . . . , not just how unlikely . . . [the SIDS explanation is]."[12] A mathematician later estimated the relative likelihood of a family's losing two babies by SIDS or by murder. He concluded, based on the available data, that two infants are 9 times more likely to be SIDS victims than murder victims.[13]

The Clarks appealed the case and, for the appeal, hired their own statisticians as expert witnesses. They lost the appeal, but they continued to seek medical explanations for the deaths and in the process uncovered the fact that the pathologist working for the prosecution had withheld the fact that the second child had been suffering from a bacterial infection at the time of death, an infection that might have caused the infant's death. Based on that discovery, a judge quashed the conviction, and after nearly three and a half years, Sally Clark was released from prison.

The renowned attorney and Harvard Law School professor Alan Dershowitz also successfully employed the prosecutor's fallacy—to help defend O. J. Simpson in his trial for the murder of Simpson's ex-wife, Nicole Brown Simpson, and a male companion. The trial of Simpson, a former football star, was one of the biggest media events of 1994–95. The police had plenty of evidence against him. They found a bloody glove at his estate that seemed to match one found at the murder scene. Bloodstains matching Nicole's blood were found on the gloves, in his white Ford Bronco, on a pair of socks in his bedroom, and in his driveway and house. Moreover, DNA samples taken from blood at the crime scene matched O. J.'s. The defense could do little more than accuse the Los Angeles Police Department of racism—O. J. is African American—and criticize the integrity of the police and the authenticity of their evidence.

The prosecution made a decision to focus the opening of its case on O. J.'s propensity toward violence against Nicole. Prosecutors spent the first ten days of the trial entering evidence of his history of abusing her and claimed that this alone was a good reason to suspect him of her murder. As they put it, "a slap is a prelude to homicide."[14] The defense attorneys used this strategy as a launchpad for their accusations of duplicity, arguing that the prosecution had spent two weeks trying to mislead the jury and that the evidence that O. J. had battered Nicole on previous occasions meant nothing. Here is Dershowitz's reasoning: 4 million women are battered annually by husbands and boyfriends in the United States, yet in 1992, according to the FBI Uniform Crime Reports, a total of 1,432, or 1 in 2,500, were killed by their husbands or boyfriends.[15] Therefore, the defense retorted, few men who slap or beat their domestic partners go on to murder them. True? Yes. Convincing? Yes. Relevant? No. The relevant number is not the probability that a man who batters his wife will go on to kill her (1 in 2,500) but rather the probability that a battered wife who was murdered was murdered by her abuser. According to the Uniform Crime Reports for the United States and Its Possessions in 1993, the probability Dershowitz (or the prosecution) should have reported was this one: of all the battered women murdered in the United States in 1993, some 90 percent were killed by their abuser. That statistic was not mentioned at the trial.

As the hour of the verdict's announcement approached, long-distance call volume dropped by half, trading volume on the New York Stock Exchange fell by 40 percent, and an estimated 100 million people turned to their televisions and radios to hear the verdict: not guilty. Dershowitz may have felt justified in misleading the jury because, in his words, "the courtroom oath—'to tell the truth, the whole truth and nothing but the truth'—is applicable only to witnesses. Defense attorneys, prosecutors, and judges don't take this oath . . . indeed, it is fair to say the American justice system is built on a foundation of *not* telling the whole truth."[16]

THOUGH CONDITIONAL PROBABILITY represented a revolution in ideas about randomness, Thomas Bayes was no revolutionary, and his work languished unattended despite its publication in the prestigious *Philosophical Transactions* in 1764. And so it fell to another man, the French scientist and mathematician Pierre-Simon de Laplace, to bring Bayes's ideas to scientists' attention and fulfill the goal of revealing to the world how the probabilities that underlie real-world situations could be inferred from the outcomes we observe.

You may remember that Bernoulli's golden theorem will tell you *before* you conduct a series of coin tosses how certain you can be, if the coin is fair, that you will observe some given outcome. You may also remember that it will not tell you *after* you've made a given series of tosses the chances that the coin was a fair one. Along the same lines, if you know that the chances that an eighty-five-year-old will survive to ninety are $^{50}/_{50}$, the golden theorem tells you the probability that half the eighty-five-year-olds in a group of 1,000 will die in the next five years, but if half the people in some group died in the five years after their eighty-fifth birthday, it cannot tell you how likely it is that the underlying chances of survival for the people in that group were $^{50}/_{50}$. Or if Ford knows that 1 in 100 of its automobiles has a defective transmission, the golden theorem can tell Ford the chances that, in a batch of 1,000 autos, 10 or more of the transmissions will be defective, but if Ford finds 10 defective transmissions in a sample of 1,000 autos, it does not tell the automaker the likelihood that the average number of defective transmissions is 1 in 100. In these cases it is the latter scenario that is more often useful in life: outside situations involving gambling, we are not normally provided with theoretical knowledge of the odds but rather must estimate them after making a series of observations. Scientists, too, find themselves in this position: they do not generally seek to know, given the value of a physical quantity, the probability that a measurement will come out one way or another but instead seek to discern the true value of a physical quantity, given a set of measurements.

I have stressed this distinction because it is an important one. It defines the fundamental difference between probability and statistics: the former concerns predictions based on fixed probabilities; the latter concerns the inference of those probabilities based on observed data.

It is the latter set of issues that was addressed by Laplace. He was not aware of Bayes's theory and therefore had to reinvent it. As he framed it, the issue was this: given a series of measurements, what is the best guess you can make of the true value of the measured quantity, and what are the chances that this guess will be "near" the true value, however demanding you are in your definition of *near*?

Laplace's analysis began with a paper in 1774 but spread over four decades. A brilliant and sometimes generous man, he also occasionally borrowed without acknowledgment from the works of others and was a tireless self-promoter. Most important, though, Laplace was a flexible reed that bent with the breeze, a characteristic that allowed him to continue his groundbreaking work virtually undisturbed by the turbulent events transpiring around him. Prior to the French Revolution, Laplace obtained the lucrative post of examiner to the royal artillery, in which he had the luck to examine a promising sixteen-year-old candidate named Napoléon Bonaparte. When the revolution came, in 1789, he fell briefly under suspicion but unlike many others emerged unscathed, declaring his "inextinguishable hatred to royalty" and eventually winning new honors from the republic. Then, when his acquaintance Napoléon crowned himself emperor in 1804, he immediately shed his republicanism and in 1806 was given the title count. After the Bourbons returned, Laplace slammed Napoléon in the 1814 edition of his treatise *Théorie analytique des probabilités*, writing that "the fall of empires which aspired to universal dominion could be predicted with very high probability by one versed in the calculus of chance."[17] The previous, 1812, edition had been dedicated to "Napoleon the Great."

Laplace's political dexterity was fortunate for mathematics, for in the end his analysis was richer and more complete than Bayes's. With the foundation provided by Laplace's work, in the next chapter we shall leave the realm of probability and enter that of statistics. Their joining point is one of the most important curves in all of mathematics and science, the bell curve, otherwise known as the normal distribution. That, and the new theory of measurement that came with it, are the subjects of the following chapter.

CHAPTER 7

Measurement and the Law of Errors

ONE DAY not long ago my son Alexei came home and announced the grade on his most recent English essay. He had received a 93. Under normal circumstances I would have congratulated him on earning an A. And since it was a low A and I know him to be capable of better, I would have added that this grade was evidence that if he put in a little effort, he could score even higher next time. But these were not normal circumstances, and in this case I considered the grade of 93 to be a shocking underestimation of the quality of the essay. At this point you might think that the previous few sentences tell you more about me than about Alexei. If so, you're right on target. In fact, the above episode is entirely about me, for it was I who wrote Alexei's essay.

Okay, shame on me. In my defense I should point out that I would normally no sooner write Alexei's essays than take a foot to the chin for him in his kung fu class. But Alexei had come to me for a critique of his work and as usual presented his request late on the night before the paper was due. I told him I'd get back to him. Proceeding to read it on the computer, I first made a couple of minor changes, nothing worth bothering to note. Then, being a relentless rewriter, I gradually found myself sucked in, rearranging this and rewriting that, and before I finished, not only had he fallen asleep, but I had made the essay my own. The next morning, sheepishly admitting that I had neglected to perform a "save as" on the original, I told him to just go ahead and turn in my version.

He handed me the graded paper with a few words of encouragement. "Not bad," he told me. "A 93 is really more of an A– than an A, but it was late and I'm sure if you were more awake, you would have done better." I was not happy. First of all, it is unpleasant when a fifteen-year-old says the very words to you that you have previously said to him, and nevertheless you find his words inane. But beyond that, how could my material—the work of a person whom my mother, at least, thinks of as a professional writer—not make the grade in a high school English class? Apparently I am not alone. Since then I have been told of another writer who had a similar experience, except his daughter received a B. Apparently the writer, with a PhD in English, writes well enough for *Rolling Stone, Esquire,* and *The New York Times* but not for English 101. Alexei tried to comfort me with another story: two of his friends, he said, once turned in identical essays. He thought that was stupid and they'd both be suspended, but not only did the overworked teacher not notice, she gave one of the essays a 90 (an A) and the other a 79 (a C). (Sounds odd unless, like me, you've had the experience of staying up all night grading a tall stack of papers with *Star Trek* reruns playing in the background to break the monotony.)

Numbers always seem to carry the weight of authority. The thinking, at least subliminally, goes like this: if a teacher awards grades on a 100-point scale, those tiny distinctions must really mean something. But if ten publishers could deem the manuscript for the first *Harry Potter* book unworthy of publication, how could poor Mrs. Finnegan (not her real name) distinguish so finely between essays as to award one a 92 and another a 93? If we accept that the quality of an essay is somehow definable, we must still recognize that a grade is not a description of an essay's degree of quality but rather a *measurement* of it, and one of the most important ways randomness affects us is through its influence on measurement. In the case of the essay the measurement apparatus was the teacher, and a teacher's assessment, like any measurement, is susceptible to random variance and error.

Voting is also a kind of measurement. In that case we are measuring not simply how many people support each candidate on election day but how many care enough to take the trouble to vote. There are many sources of random error in this measurement. Some legitimate voters might find that their name is not on the rolls of registered voters. Others mistakenly vote for a candidate other than the one intended. And of course there are errors in counting the votes. Some ballots are improperly accepted or rejected; others are simply lost. In most elections the sum of all these factors doesn't add up to enough to affect the outcome. But in close elections it can, and then we usually go through one or more recounts, as if our second or third counting of the votes will be less affected by random errors than our first.

In the 2004 governor's race in the state of Washington, for example, the Democratic candidate was eventually declared the winner although the original tally had the Republican winning by 261 votes out of about 3 million.[1] Since the original vote count was so close, state law required a recount. In that count the Republican won again, but by only 42 votes. It is not known whether anyone thought it was a bad sign that the 219-vote difference between the first and second vote counts was several times larger than the new margin of victory, but the upshot was a third vote count, this one entirely "by hand." The 42-vote victory amounted to an edge of just 1 vote out of each 70,000 cast, so the hand-counting effort could be compared to asking 42 people to count from 1 to 70,000 and then hoping they averaged less than 1 mistake each. Not surprisingly, the result changed again. This time it favored the Democrat by 10 votes. That number was later changed to 129 when 700 newly discovered "lost votes" were included.

Neither the vote-counting process nor the voting process is perfect. If, for instance, owing to post office mistakes, 1 in 100 prospective voters didn't get the mailer with the location of the polling place and 1 in 100 of those people did not vote because of it, in the Washington election that would have amounted to 300 voters who would have voted but didn't because of government error. Elections, like all measurements, are imprecise, and so are the recounts, so when elections come out extremely close, perhaps we ought to accept them as is, or flip a coin, rather than conducting recount after recount.

The imprecision of measurement became a major issue in the mid-eighteenth century, when one of the primary occupations of those working in celestial physics and mathematics was the problem of reconciling Newton's laws with the observed motions of the moon and planets. One way to produce a single number from a set of discordant measurements is to take the average, or mean. It seems to have been young Isaac Newton who, in his optical investigations, first employed it for that purpose.[2] But as in many things, Newton was an anomaly. Most scientists in Newton's day, and in the following century, didn't take the mean. Instead, they chose the single "golden number" from among their measurements—the number they deemed mainly by hunch to be the most reliable result they had. That's because they regarded variation in measurement not as the inevitable by-product of the measuring process but as evidence of failure—with, at times, even moral consequences. In fact, they rarely published multiple measurements of the same quantity, feeling it would amount to the admission of a botched process and raise the issue of trust. But in the mid-eighteenth century the tide began to change. Calculating the gross motion of heavenly bodies, a series of nearly circular ellipses, is a simple task performed today by precocious high school students as music blares through their headphones. But to describe planetary motion in its finer points, taking into account not only the gravitational pull of the sun but also that of the other planets and the deviation of the planets and the moon from a perfectly spherical shape, is even today a difficult problem. To accomplish that goal, complex and approximate mathematics had to be reconciled with imperfect observation and measurement.

There was another reason why the late eighteenth century demanded a mathematical theory of measurement: beginning in the 1780s in France a new mode of rigorous experimental physics had arisen.[3] Before that period, physics consisted of two separate traditions. On the one hand, mathematical scientists investigated the precise consequences of Newton's theories of motion and gravity. On the other, a group sometimes described as experimental philosophers performed empirical investigations of electricity, magnetism, light, and heat. The experimental philosophers—often amateurs—were less focused on the rigorous methodology of science than were the mathematics-oriented researchers, and so a movement arose to reform and mathematize experimental physics. In it Pierre-Simon de Laplace again played a major role.

Laplace had become interested in physical science through the work of his fellow Frenchman Antoine-Laurent Lavoisier, considered the father of modern chemistry.[4] Laplace and Lavoisier worked together for years, but Lavoisier did not prove as adept as Laplace at navigating the troubled times. To earn money to finance his many scientific experiments, he had become a member of a privileged private association of state-protected tax collectors. There is probably no time in history when having such a position would inspire your fellow citizens to invite you into their homes for a nice hot cup of gingerbread cappuccino, but when the French Revolution came, it proved an especially onerous credential. In 1794, Lavoisier was arrested with the rest of the association and quickly sentenced to death. Ever the dedicated scientist, he requested time to complete some of his research so that it would be available to posterity. To that the presiding judge famously replied, "The republic has no need of scientists." The father of modern chemistry was promptly beheaded, his body tossed into a mass grave. He had reportedly instructed his assistant to count the number of words his severed head would attempt to mouth.

Laplace's and Lavoisier's work, along with that of a few others, especially the French physicist Charles-Augustin de Coulomb, who experimented on electricity and magnetism, transformed experimental physics. Their work also contributed to the development, in the 1790s, of a new rational system of units, the metric system, to replace the disparate systems that had impeded science and were a frequent cause of dispute among merchants. Developed by a group appointed by Louis XVI, the metric system was adopted by the revolutionary government after Louis's downfall. Lavoisier, ironically, had been one of the group's members.

The demands of both astronomy and experimental physics meant that a great part of the mathematician's task in the late eighteenth and early nineteenth centuries was understanding and quantifying random error. Those efforts led to a new field, mathematical statistics, which provides a set of tools for the interpretation of the data that arise from observation and experimentation. Statisticians sometimes view the growth of modern science as revolving around that development, the creation of a theory of measurement. But statistics also provides tools to address real-world issues, such as the effectiveness of drugs or the popularity of politicians, so a proper understanding of statistical reasoning is as useful in everyday life as it is in science.

IT IS ONE OF THOSE CONTRADICTIONS of life that although measurement always carries uncertainty, the uncertainty in measurement is rarely discussed when measurements are quoted. If a fastidious traffic cop tells the judge her radar gun clocked you going thirty-nine in a thirty-five-mile-per-hour zone, the ticket will usually stick despite the fact that readings from radar guns often vary by several miles per hour.[5] And though many students (along with their parents) would jump off the roof if doing so would raise their 598 on the math SAT to a 625, few educators talk about the studies showing that, if you want to gain 30 points, there's a good chance you can do it simply by taking the test a couple more times.[6] Sometimes meaningless distinctions even make the news. One recent August the Bureau of Labor Statistics reported that the unemployment rate stood at 4.7 percent. In July the bureau had reported the rate at 4.8 percent. The change prompted headlines like this one in *The New York Times:* "Jobs and Wages Increased Modestly Last Month."[7] But as Gene Epstein, the economics editor of *Barron's*, put it, "Merely because the number has changed it doesn't necessarily mean that a thing itself has changed. For example, any time the unemployment rate moves by a tenth of a percentage point . . . that is a change that is so small, there is no way to tell whether there really was a change."[8] In other words, if the Bureau of Labor Statistics measures the unemployment rate in August and then repeats its measurement an hour later, by random error alone there is a good chance that the second measurement will differ from the first by at least a tenth of a percentage point. Would *The New York Times* then run the headline "Jobs and Wages Increased Modestly at 2 P.M."?

The uncertainty in measurement is even more problematic when the quantity being measured is subjective, like Alexei's English-class essay. For instance, a group of researchers at Clarion University of Pennsylvania collected 120 term papers and treated them with a degree of scrutiny you can be certain your own child's work will never receive: each term paper was scored independently by eight faculty members. The resulting grades, on a scale from A to F, sometimes varied by two or more grades. On average they differed by nearly one grade.[9] Since a student's future often depends on such judgments, the imprecision is unfortunate. Yet it is understandable given that, in their approach and philosophy, the professors in any given college department often run the gamut from Karl Marx to Groucho Marx. But what if we control for that—that is, if the graders are given, and instructed to follow, certain fixed grading criteria? A researcher at Iowa State University presented about 100 students' essays to a group of doctoral students in rhetoric and professional communication whom he had trained extensively according to such criteria.[10] Two independent assessors graded each essay on a scale of 1 to 4. When the scores were compared, the assessors agreed in only about half the cases. Similar results were found by the University of Texas in an analysis of its scores on college-entrance essays.[11] Even the venerable College Board expects only that, when assessed by two raters, "92% of all scored essays will receive ratings within ± 1 point of each other on the 6-point SAT essay scale."[12]

Another subjective measurement that is given more credence

than it warrants is the rating of wines. Back in the 1970s the wine business was a sleepy enterprise, growing, but mainly in the sales of low-grade jug wines. Then, in 1978, an event often credited with the rapid growth of that industry occurred: a lawyer turned self-proclaimed wine critic, Robert M. Parker Jr., decided that, in addition to his reviews, he would rate wines numerically on a 100-point scale. Over the years most other wine publications followed suit. Today annual wine sales in the United States exceed $20 billion, and millions of wine aficionados won't lay their money on the counter without first looking to a wine's rating to support their choice. So when *Wine Spectator* awarded, say, the 2004 Valentín Bianchi Argentine cabernet sauvignon a 90 rather than an 89, that single extra point translated into a huge difference in Valentín Bianchi's sales.[13] In fact, if you look in your local wine shop, you'll find that the sale and bargain wines, owing to their lesser appeal, are often the wines rated in the high 80s. But what are the chances that the 2004 Valentín Bianchi Argentine cabernet that received a 90 would have received an 89 if the rating process had been repeated, say, an hour later?

In his 1890 book *The Principles of Psychology*, William James suggested that wine expertise could extend to the ability to judge whether a sample of Madeira came from the top or the bottom of a bottle.[14] In the wine tastings that I've attended over the years, I've noticed that if the bearded fellow to my left mutters "a great nose" (the wine smells good), others certainly might chime in their agreement. But if you make your notes independently and without discussion, you often find that the bearded fellow wrote, "Great nose"; the guy with the shaved head scribbled, "No nose"; and the blond woman with the perm wrote, "Interesting nose with hints of parsley and freshly tanned leather."

From the theoretical viewpoint, there are many reasons to question the significance of wine ratings. For one thing, taste perception depends on a complex interaction between taste and olfactory stimulation. Strictly speaking, the sense of taste comes from five types of receptor cells on the tongue: salty, sweet, sour, bitter, and umami. The last responds to certain amino acid compounds (prevalent, for

example, in soy sauce). But if that were all there was to taste perception, you could mimic everything—your favorite steak, baked potato,
and apple pie feast or a nice spaghetti Bolognese—employing only
table salt, sugar, vinegar, quinine, and monosodium glutamate. Fortunately there is more to gluttony than that, and that is where the
sense of smell comes in. The sense of smell explains why, if you take
two identical solutions of sugar water and add to one a (sugar-free)
essence of strawberry, it will taste sweeter than the other.[15] The perceived taste of wine arises from the effects of a stew of between 600
and 800 volatile organic compounds on both the tongue and the
nose.[16] That's a problem, given that studies have shown that even
flavor-trained professionals can rarely reliably identify more than
three or four components in a mixture.[17]

Expectations also affect your perception of taste. In 1963 three
researchers secretly added a bit of red food color to white wine to give
it the blush of a rosé. They then asked a group of experts to rate its
sweetness in comparison with the untinted wine. The experts perceived the fake rosé as sweeter than the white, according to their
expectation. Another group of researchers gave a group of oenology
students two wine samples. Both samples contained the same white
wine, but to one was added a tasteless grape anthocyanin dye that
made it appear to be red wine. The students also perceived differences between the red and the white corresponding to their expectations.[18] And in a 2008 study a group of volunteers asked to rate five
wines rated a bottle labeled $90 higher than another bottle labeled
$10, even though the sneaky researchers had filled both bottles with
the same wine. What's more, this test was conducted while the subjects were having their brains imaged in a magnetic resonance scan-
ner. The scans showed that the area of the brain thought to encode
our experience of pleasure was truly more active when the subjects
drank the wine they believed was more expensive.[19] But before you
judge the oenophiles, consider this: when a researcher asked 30 cola
drinkers whether they preferred Coke or Pepsi and then asked them
to test their preference by tasting both brands side by side, 21 of the
30 reported that the taste test confirmed their choice even though

this sneaky researcher had put Coke in the Pepsi bottle and vice versa.[20] When we perform an assessment or measurement, our brains do not rely solely on direct perceptional input. They also integrate other sources of information—such as our expectation.

Wine tasters are also often fooled by the flip side of the expectancy bias: a lack of context. Holding a chunk of horseradish under your nostril, you'd probably not mistake it for a clove of garlic, nor would you mistake a clove of garlic for, say, the inside of your sneaker. But if you sniff clear liquid scents, all bets are off. In the absence of context, there's a good chance you'd mix the scents up. At least that's what happened when two researchers presented experts with a series of sixteen random odors: the experts misidentified about 1 out of every 4 scents.[21]

Given all these reasons for skepticism, scientists designed ways to measure wine experts' taste discrimination directly. One method is to use a wine triangle. It is not a physical triangle but a metaphor: each expert is given three wines, two of which are identical. The mission: to choose the odd sample. In a 1990 study, the experts identified the odd sample only two-thirds of the time, which means that in 1 out of 3 taste challenges these wine gurus couldn't distinguish a pinot noir with, say, "an exuberant nose of wild strawberry, luscious black-berry, and raspberry," from one with "the scent of distinctive dried plums, yellow cherries, and silky cassis."[22] In the same study an ensemble of experts was asked to rank a series of wines based on 12 components, such as alcohol content, the presence of tannins, sweetness, and fruitiness. The experts disagreed significantly on 9 of the 12 components. Finally, when asked to match wines with the descriptions provided by other experts, the subjects were correct only 70 percent of the time.

Wine critics are conscious of all these difficulties. "On many levels . . . [the ratings system] is nonsensical," says the editor of *Wine and Spirits Magazine*.[23] And according to a former editor of *Wine Enthusiast*, "The deeper you get into this the more you realize how misguided and misleading this all is."[24] Yet the rating system thrives. Why? The critics found that when they attempted to encapsulate wine quality with a system of stars or simple verbal descriptors such as *good, bad,* and maybe *ugly,* their opinions were unconvincing. But when they used numbers, shoppers worshipped their pronouncements. Numerical ratings, though dubious, make buyers confident that they can pick the golden needle (or the silver one, depending on their budget) from the haystack of wine varieties, makers, and vintages.

If a wine—or an essay—truly admits some measure of quality that can be summarized by a number, a theory of measurement must address two key issues: How do we determine that number from a series of varying measurements? And given a limited set of measurements, how can we assess the probability that our determination is correct? We now turn to these questions, for whether the source of data is objective or subjective, their answers are the goal of the theory of measurement.

THE KEY to understanding measurement is understanding the nature of the variation in data caused by random error. Suppose we offer a number of wines to fifteen critics or we offer the wines to one critic repeatedly on different days or we do both. We can neatly summarize the opinions employing the average, or mean, of the ratings. But it is not just the mean that matters: if all fifteen critics agree that the wine is a 90, that sends one message; if the critics produce the ratings 80, 81, 82, 87, 89, 89, 90, 90, 90, 91, 91, 94, 97, 99, and 100, that sends another. Both sets of data have the same mean, but they differ in the amount they vary from that mean. Since the manner in which data points are distributed is such an important piece of information, mathematicians created a numerical measure of variation to describe it. That number is called the sample standard deviation. Mathematicians also measure the variation by its square, which is called the sample variance.

The sample standard deviation characterizes how close to the mean a set of data clusters or, in practical terms, the uncertainty of the data. When it is low, the data fall near the mean. For the data in which all wine critics rated the wine 90, for example, the sample standard deviation is 0, telling you that all the data are identical to the mean. When the sample standard deviation is high, however, the data are not clustered around the mean. For the set of wine ratings above that ranges from 80 to 100, the sample standard deviation is 6, meaning that as a rule of thumb most of the ratings fall within 6 points of the mean. In that case all you can really say about the wine is that it is probably somewhere between an 84 and a 96.

In judging the meaning of their measurements, scientists in the eighteenth and nineteenth centuries faced the same issues as the skeptical oenophile. For if a group of researchers makes a series of observations, the results will almost always differ. One astronomer might suffer adverse atmospheric conditions; another might be jostled by a breeze; a third might have just returned from a Madeira tasting with William James. In 1838 the mathematician and astronomer F. W. Bessel categorized eleven classes of random errors that occur in every telescopic observation. Even if a single astronomer makes repeated measurements, variables such as unreliable eyesight or the effect of temperature on the apparatus will cause the observations to vary. And so astronomers must understand how, given a series of discrepant measurements, they can determine a body's true position. But just because oenophiles and scientists share a problem, it doesn't mean they can share its solution. Can we identify general characteristics of random error, or does the character of random error depend on the context?

One of the first to imply that diverse sets of measurements share common characteristics was Jakob Bernoulli's nephew Daniel. In 1777 he likened the random errors in astronomical observation to the deviations in the flight of an archer's arrows. In both cases, he reasoned, the target—true value of the measured quantity, or the bull's-eye—should lie somewhere near the center, and the observed results should be bunched around it, with more reaching the inner bands and fewer falling farther from the mark. The law he proposed to describe the distribution did not prove to be the correct one, but what is important is the insight that the distribution of an archer's errors might mirror the distribution of errors in astronomical observations.

That the distribution of errors follows some universal law, sometimes called the error law, is the central precept on which the theory of measurement is based. Its magical implication is that, given that certain very common conditions are satisfied, any determination of a true value based on measured values can be solved employing a single mathematical analysis. When such a universal law is employed, the problem of determining the true position of a heavenly body based on a set of astronomers' measurements is equivalent to that of determining the position of a bull's-eye given only the arrow holes or a wine's "quality" given a series of ratings. That is the reason mathematical statistics is a coherent subject rather than merely a bag of tricks: whether your repeated measurements are aimed at determining the position of Jupiter at 4 A.M. on Christmas Day or the weight of a loaf of raisin bread coming off an assembly line, the distribution of errors is the same.

This doesn't mean random error is the only kind of error that can affect measurement. If half a group of wine critics liked only red wines and the other half only white wines but they all otherwise agreed perfectly (and were perfectly consistent), then the ratings earned by a particular wine would not follow the error law but instead would consist of two sharp peaks, one due to the red wine lovers and one due to the white wine lovers. But even in situations where the applicability of the law may not be obvious, from the point spreads of pro football games[25] to IQ ratings, the error law often does apply. Many years ago I got hold of a few thousand registration cards for a consumer software program a friend had designed for eight- and nine-year-olds. The software wasn't selling as well as expected. Who was buying it? After some tabulation I found that the greatest number of users occurred at age seven, indicating an unwelcome but not unexpected mismatch. But what was truly striking was that when I made a bar graph showing how the number of buyers diminished as the buyers' age strayed from the mean of seven, I found that the graph took a very familiar shape—that of the error law.

It is one thing to suspect that archers and astronomers, chemists and marketers, encounter the same error law; it is another to discover the specific form of that law. Driven by the need to analyze astronomical data, scientists like Daniel Bernoulli and Laplace postulated a series of flawed candidates in the late eighteenth century. As it turned out, the correct mathematical function describing the error law—the bell curve—had been under their noses the whole time. It had been discovered in London in a different context many decades earlier.

OF THE THREE PEOPLE instrumental in uncovering the importance of the bell curve, its discoverer is the one who least often gets the credit. Abraham De Moivre's breakthrough came in 1733, when he was in his mid-sixties, and wasn't made public until his book *The Doctrine of Chances* came out in its second edition five years later. De Moivre was led to the curve while searching for an approximation to the numbers that inhabit the regions of Pascal's triangle far beneath the place where I truncated it, hundreds or thousands of lines down. In order to prove his version of the law of large numbers, Jakob Bernoulli had had to grapple with certain properties of the numbers that appeared in those lines. The numbers can be very large—for instance, one coefficient in the 200th row of Pascal's triangle has fifty-nine digits! In Bernoulli's day, and indeed in the days before computers, such numbers were obviously very hard to calculate. That's why, as I said, Bernoulli proved his law of large numbers employing various approximations, which diminished the practical usefulness of his result. With his curve, De Moivre was able to make far better approximations to the coefficients and therefore greatly improve on Bernoulli's estimates.

The approximation De Moivre derived is evident if, as I did for the registration cards, you represent the numbers in a row of the triangle by the height of the bars on a bar graph. For instance, the three numbers in the third line of the triangle are 1, 2, 1. In their bar graph the first bar rises one unit; the second is twice that height; and the third is again just one unit. Now look at the five numbers in the fifth line: 1, 4, 6, 4, 1. That graph will have five bars, again starting low, rising to a peak at the center, and then falling off symmetrically. The coefficients very far down in the triangle lead to bar graphs with very many bars, but they behave in the same manner. The bar graphs in the case of the 10th, 100th, and 1,000th lines of Pascal's triangle are shown on page 139.

If you draw a curve connecting the tops of all the bars in each bar graph, it will take on a characteristic shape, a shape approaching that of a bell. And if you smooth the curve a bit, you can write a mathematical expression for it. That smooth bell curve is more than just a visualization of the numbers in Pascal's triangle; it is a means for obtaining an accurate and easy-to-use estimate of the numbers that appear in the triangle's lower lines. This was De Moivre's discovery.

Today the bell curve is usually called the normal distribution and sometimes the Gaussian distribution (we'll see later where that term originated). The normal distribution is actually not a fixed curve but a family of curves, in which each depends on two parameters to set its specific position and shape. The first parameter determines where its peak is located, which is at 5, 50, and 500 in the graphs on page 139. The second parameter determines the amount of spread in the curve. Though it didn't receive its modern name until 1894, this measure is called the standard deviation, and it is the theoretical counterpart of the concept I spoke of earlier, the sample standard deviation. Roughly speaking, it is half the width of the curve at the point at which the curve is about 60 percent of its maximum height. Today the importance of the normal distribution stretches far beyond its use as an approximation to the numbers in Pascal's triangle. It is, in fact, the most widespread manner in which data have been found to be distributed.

When employed to describe the distribution of data, the bell curve describes how, when you make many observations, most of them fall around the mean, which is represented by the peak of the

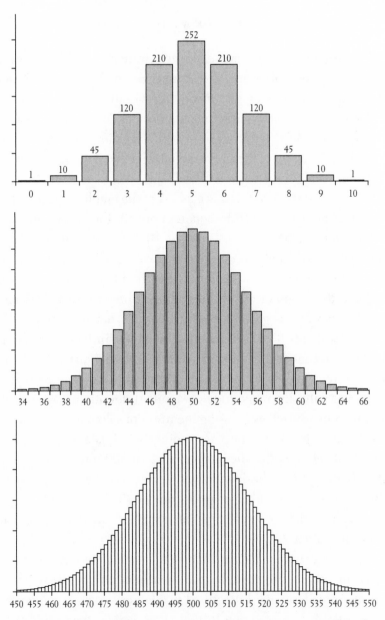

The bars in the graphs above represent the relative magnitudes of the entries in the 10th, 100th, and 1,000th rows of Pascal's triangle (see page 72). The numbers along the horizontal axis indicate to which entry the bar refers. By convention, that labeling begins at 0, rather than 1 (the middle and bottom graphs have been truncated so that the entries whose bars would have negligible height are not shown).

139

curve. Moreover, as the curve slopes symmetrically downward on either side, it describes how the number of observations diminishes equally above and below the mean, at first rather sharply and then less drastically. In data that follow the normal distribution, about 68 percent (roughly two-thirds) of your observations will fall within 1 standard deviation of the mean, about 95 percent within 2 standard deviations, and 99.7 percent within 3.

In order to visualize this, have a look at the graph on page 141. In this table the data marked by squares concern the guesses made by 300 students, each observing a series of 10 coin flips.[26] Along the horizontal axis is plotted the number of correct guesses, from 0 to 10. Along the vertical axis is plotted the number of students who achieved that number of correct guesses. The curve is bell shaped, centered at 5 correct guesses, at which point its height corresponds to about 75 students. The curve falls to about two-thirds of its maximum height, corresponding to about 51 students, about halfway between 3 and 4 correct guesses on the left and between 6 and 7 on the right. A bell curve with this magnitude of standard deviation is typical of a random process such as guessing the result of a coin toss.

The same graph also displays another set of data, marked by circles. That set describes the performance of 300 mutual fund managers. In this case the horizontal axis represents not correct guesses of coin flips but the number of years (out of 10) that a manager performed above the group average. Note the similarity! We'll get back to this in chapter 9.

A good way to get a feeling for how the normal distribution relates to random error is to consider the process of polling, or sampling. You may recall the poll I described in chapter 5 regarding the popularity of the mayor of Basel. In that city a certain fraction of voters approved of the mayor, and a certain fraction disapproved. For the sake of simplicity we will now assume each was 50 percent. As we saw, there is a chance that those involved in the poll would not reflect exactly this 50/50 split. In fact, if N voters were questioned, the chances that any given number of them would support the mayor are proportional to the numbers on line N of Pascal's triangle. And so,

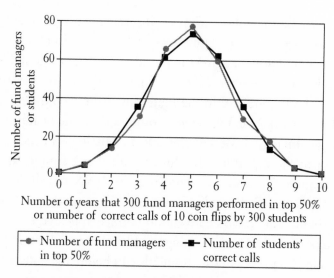

Number of years that 300 fund managers performed in top 50%
or number of correct calls of 10 coin flips by 300 students

| —●— Number of fund managers in top 50% | —■— Number of students' correct calls |

Coin toss guessing compared to stock-picking success

according to De Moivre's work, if pollsters poll a large number of voters, the probabilities of different polling results can be described by the normal distribution. In other words about 95 percent of the time the approval rating they observe in their poll will fall within 2 standard deviations of the true rating, 50 percent. Pollsters use the term *margin of error* to describe this uncertainty. When pollsters tell the media that a poll's margin of error is plus or minus 5 percent, they mean that if they were to repeat the poll a large number of times, 19 out of 20 (95 percent) of those times the result would be within 5 percent of the correct answer. (Though pollsters rarely point this out, that also means, of course, that about 1 time in 20 the result will be wildly inaccurate.) As a rule of thumb, a sample of 100 yields a margin of error that is too great for most purposes. A sample of 1,000, on the other hand, usually yields a margin of error in the ballpark of 3 percent, which for most purposes suffices.

It is important, whenever assessing any kind of survey or poll, to realize that when it is repeated, we should expect the results to vary. For example, if in reality 40 percent of registered voters approve of the way the president is handling his job, it is much more likely that six independent surveys will report numbers like 37, 39, 39, 40, 42,

and 42 than it is that all six surveys will agree that the president's support stands at 40 percent. (Those six numbers are in fact the results of
six independent polls gauging the president's job approval in the first
two weeks of September 2006.)[27] That's why, as another rule of
thumb, any variation within the margin of error should be ignored.
But although *The New York Times* would not run the headline "Jobs
and Wages Increased Modestly at 2 P.M.," analogous headlines are
common in the reporting of political polls. For example, after the
Republican National Convention in 2004, CNN ran the headline
"Bush Apparently Gets Modest Bounce."[28] The experts at CNN went
on to explain that "Bush's convention bounce appeared to be 2 percentage points. . . . The percentage of likely voters who said he was
their choice for president rose from 50 right before the convention to
52 immediately afterward." Only later did the reporter remark that
the poll's margin of error was plus or minus 3.5 percentage points,
which means that the news flash was essentially meaningless. Apparently the word *apparently*, in CNN-talk, means "apparently not."

For many polls a margin of error of more than 5 percent is considered unacceptable, yet in our everyday lives we make judgments
based on far fewer data points than that. People don't get to play 100
years of professional basketball, invest in 100 apartment buildings, or
start 100 chocolate-chip-cookie companies. And so when we judge
their success at those enterprises, we judge them on just a few data
points. Should a football team lavish $50 million to lure a guy com-
ing off a single record-breaking year? How likely is it that the stockbroker who wants your money for a sure thing will repeat her earlier
successes? Does the success of the wealthy inventor of sea monkeys
mean there is a good chance he'll succeed with his new ideas of invisible goldfish and instant frogs? (For the record, he didn't.)[29] When
we observe a success or a failure, we are observing one data point, a
sample from under the bell curve that represents the potentialities
that previously existed. We cannot know whether our single observation represents the mean or an outlier, an event to bet on or a rare
happening that is not likely to be reproduced. But at a minimum we
ought to be aware that a sample point is just a sample point, and

rather than accepting it simply as reality, we ought to see it in the context of the standard deviation or the spread of possibilities that produced it. The wine might be rated 91, but that number is meaningless if we have no estimate of the variation that would occur if the identical wine were rated again and again or by someone else. It might help to know, for instance, that a few years back, when both *The Penguin Good Australian Wine Guide* and On Wine's *Australian Wine Annual* reviewed the 1999 vintage of the Mitchelton Blackwood Park Riesling, the *Penguin* guide gave the wine five stars out of five and named it Penguin Best Wine of the Year, while On Wine rated it at the bottom of all the wines it reviewed, deeming it the worst vintage produced in a decade.[30] The normal distribution not only helps us understand such discrepancies, but also has enabled a myriad of statistical applications widely employed today in both science and commerce—for example, whenever a drug company assesses whether the results of a clinical trial are significant, a manufacturer assesses whether a sample of parts accurately reflects the proportion of those that are defective, or a marketer decides whether to act on the results of a research survey.

THE RECOGNITION that the normal distribution describes the distribution of measurement error came decades after De Moivre's work, by that fellow whose name is sometimes attached to the bell curve, the German mathematician Carl Friedrich Gauss. It was while working on the problem of planetary motion that Gauss came to that realization, at least regarding astronomical measurements. Gauss's "proof," however, was, by his own later admission, invalid.[31] Moreover, its far-reaching consequences also eluded him. And so he slipped the law inconspicuously into a section at the end of a book called *The Theory of the Motion of Heavenly Bodies Moving about the Sun in Conic Sections.* There it may well have died, just another in the growing pile of abandoned proposals for the error law.

It was Laplace who plucked the normal distribution from obscurity. He encountered Gauss's work in 1810, soon after he had read a memoir to the Académie des Sciences proving a theorem called the central limit theorem, which says that the probability that the sum of a large number of independent random factors will take on any given value is distributed according to the normal distribution. For example, suppose you bake 100 loaves of bread, each time following a recipe that is meant to produce a loaf weighing 1,000 grams. By chance you will sometimes add a bit more or a bit less flour or milk, or a bit more or less moisture may escape in the oven. If in the end each of a myriad of possible causes adds or subtracts a few grams, the central limit theorem says that the weight of your loaves will vary according to the normal distribution. Upon reading Gauss's work, Laplace immediately realized that he could use it to improve his own and that his work could provide a better argument than Gauss's to support the notion that the normal distribution is indeed the error law. Laplace rushed to press a short sequel to his memoir on the theorem. Today the central limit theorem and the law of large numbers are the two most famous results of the theory of randomness.

To illustrate how the central limit theorem explains why the normal distribution is the correct error law, let's reconsider Daniel Bernoulli's example of the archer. I played the role of the archer one night after a pleasant interlude of wine and adult company, when my younger son, Nicolai, handed me a bow and arrow and dared me to shoot an apple off his head. The arrow had a soft foam tip, but still it seemed reasonable to conduct an analysis of my possible errors and their likelihood. For obvious reasons I was mainly concerned with vertical errors. A simple model of the errors is this: Each random factor—say, a sighting error, the effect of air currents, and so on—would throw my shot vertically off target, either high or low, with equal probability. My total error in aim would then be the sum of my errors. If I was lucky, about half the component errors would deflect the arrow upward and half downward, and my shot would end up right on target. If I was unlucky (or, more to the point, if my son was unlucky), the errors would all fall one way and my aim would be far off, either high or low. The relevant question was, how likely was it that the errors would cancel each other, or that they would add up to their maximum, or that they would take any other value in between? But that was just a Bernoulli process—like tossing coins and asking how likely it is that the tosses will result in a certain number of heads. The answer is described by Pascal's triangle or, if many trials are involved, by the normal distribution. And that, in this case, is precisely what the central limit theorem tells us. (As it turned out, I missed both apple and son, but did knock over a glass of very nice cabernet.)

By the 1830s most scientists had come to believe that every measurement is a composite, subject to a great number of sources of deviation and hence to the error law. The error law and the central limit theorem thus allowed for a new and deeper understanding of data and their relation to physical reality. In the ensuing century, scholars interested in human society also grasped these ideas and found to their surprise that the variation in human characteristics and behavior often displays the same pattern as the error in measurement. And so they sought to extend the application of the error law from physical science to a new science of human affairs.

CHAPTER 8

The Order in Chaos

I N THE MID-1960S, some ninety years old and in great need of money to live on, a Frenchwoman named Jeanne Calment made a deal with a forty-seven-year-old lawyer: she sold him her apartment for the price of a low monthly subsistence payment with the agreement that the payments would stop upon her death, at which point she would be carried out and he could move in.[1] The lawyer must have known that Ms. Calment had already exceeded the French life expectancy by more than ten years. He may not have been aware of Bayes's theory, however, nor known that the relevant issue was not whether she should be expected to die in minus ten years but that her life expectancy, given that she had already made it to ninety, was about six more years.[2] Still, he had to feel comfortable believing that any woman who as a teenager had met Vincent van Gogh in her father's shop would soon be joining van Gogh in the hereafter. (For the record, she found the artist "dirty, badly dressed, and disagreeable.")

Ten years later the attorney had presumably found an alternative dwelling, for Jeanne Calment celebrated her 100th birthday in good health. And though her life expectancy was then about two years, she reached her 110th birthday still on the lawyer's dime. By that point the attorney had turned sixty-seven. But it was another decade before the attorney's long wait came to an end, and it wasn't the end he expected. In 1995 the attorney himself died while Jeanne Calment

146

lived on. Her day of reckoning finally came on August 4, 1997, at the age of 122. Her age at death exceeded the lawyer's age at his death by forty-five years.

Individual life spans—and lives—are unpredictable, but when data are collected from groups and analyzed en masse, regular patterns emerge. Suppose you have driven accident-free for twenty years. Then one fateful afternoon while you're on vacation in Quebec with your spouse and your in-laws, your mother-in-law yells, "Look out for that moose!" and you swerve into a warning sign that says essentially the same thing. To you the incident would feel like an odd and unique event. But as the need for the sign indicates, in an ensemble of thousands of drivers a certain percentage of drivers can be counted on to encounter a moose. In fact, a statistical ensemble of people acting randomly often displays behavior as consistent and predictable as a group of people pursuing conscious goals. Or as the philosopher Immanuel Kant wrote in 1784, "Each, according to his own inclination follows his own purpose, often in opposition to others; yet each individual and people, as if following some guiding thread, go toward a natural but to each of them unknown goal; all work toward furthering it, even if they would set little store by it if they did know it."[3]

According to the Federal Highway Administration, for example, there are about 200 million drivers in the United States.[4] And according to the National Highway Traffic Safety Administration, in one recent year those drivers drove a total of about 2.86 trillion miles.[5] That's about 14,300 miles per driver. Now suppose everyone in the country had decided it would be fun to hit that same total again the following year. Let's compare two methods that could have been used to achieve that goal. In method 1 the government institutes a rationing system employing one of the National Science Foundation's supercomputing centers to assign personal mileage targets that meet each of the 200 million motorists' needs while maintaining the previous annual average of 14,300. In method 2 we tell drivers not to stress out over it and to drive as much or as little as they please with no regard to how far they drove the prior year. If Uncle Billy

Bob, who used to walk to work at the liquor store, decides instead to log 100,000 miles as a shotgun wholesaler in West Texas, that's fine. And if Cousin Jane in Manhattan, who logged most of her mileage circling the block on street-cleaning days in search of a parking space, gets married and moves to New Jersey, we won't worry about that either. Which method would come closer to the target of 14,300 miles per driver? Method 1 is impossible to test, though our limited experience with gasoline rationing indicates that it probably wouldn't work very well. Method 2, on the other hand, was actually instituted—that is, the following year, drivers drove as much or as little as they pleased without attempting to hit any quota. How did they do? According to the National Highway Traffic Safety Administration, that year American drivers drove 2.88 trillion miles, or 14,400 miles per person, only 100 miles above target. What's more, those 200 million drivers also suffered, within less than 200, the same number of fatalities in both years (42,815 versus 42,643).

We associate randomness with disorder. Yet although the lives of 200 million drivers vary unforeseeably, in the aggregate their behavior could hardly have proved more orderly. Analogous regularities can be found if we examine how people vote, buy stocks, marry, are told to get lost, misaddress letters, or sit in traffic on their way to a meeting they didn't want to go to in the first place—or if we measure the length of their legs, the size of their feet, the width of their buttocks, or the breadth of their beer bellies. As nineteenth-century scientists dug into newly available social data, wherever they looked, the chaos of life seemed to produce quantifiable and predictable patterns. But it was not just the regularities that astonished them. It was also the nature of the variation. Social data, they discovered, often follow the normal distribution.

That the variation in human characteristics and behavior is distributed like the error in an archer's aim led some nineteenth-century scientists to study the targets toward which the arrows of human existence are aimed. More important, they sought to understand the social and physical causes that sometimes move the target. And so the field of mathematical statistics, developed to aid scientists in data analysis, flourished in a far different realm: the study of the nature of society.

STATISTICIANS have been analyzing life's data at least since the eleventh century, when William the Conqueror commissioned what was, in effect, the first national census. William began his rule in 1035, at age seven, when he succeeded his father as duke of Normandy. As his moniker implies, Duke William II liked to conquer, and in 1066 he invaded England. By Christmas Day he was able to give himself the present of being crowned king. His swift victory left him with a little problem: whom exactly had he conquered, and more important, how much could he tax his new subjects? To answer those questions, he sent inspectors into every part of England to note the size, ownership, and resources of each parcel of land.[6] To make sure they got it right, he sent a second set of inspectors to duplicate the effort of the first set. Since taxation was based not on population but on land and its usage, the inspectors made a valiant effort to count every ox, cow, and pig but didn't gather much data about the people who shoveled their droppings. Even if population data had been relevant, in medieval times a statistical survey regarding the most vital statistics about humans—their life spans and diseases— would have been considered inconsistent with the traditional Christian concept of death. According to that doctrine, it was wrong to make death the object of speculation and almost sacrilegious to look for laws governing it. For whether a person died from a lung infection, a stomachache, or a rock whose impact exceeded the compressive strength of his skull, the true cause of his or her death was considered to be simply God's will. Over the centuries that fatalistic attitude gradually gave way, yielding to an opposing view, according to which, by studying the regularities of nature and society, we are not challenging God's authority but rather learning about his ways.

A big step in that transformation of views came in the sixteenth century, when the lord mayor of London ordered the compilation of weekly "bills of mortality" to account for the christenings and burials

recorded by parish clerks. For decades the bills were compiled spo-
radically, but in 1603, one of the worst years of the plague, the city
instituted a weekly tally. Theorists on the Continent turned up their
noses at the data-laden mortality bills as peculiarly English and of little use. But to one peculiar Englishman, a shopkeeper named John
Graunt, the tallies told a gripping tale.[7]

Graunt and his friend William Petty have been called the
founders of statistics, a field sometimes considered lowbrow by those
in pure mathematics owing to its focus on mundane practical issues,
and in that sense Graunt in particular makes a fitting founder. For
unlike some of the amateurs who developed probability—Cardano
the doctor, Fermat the jurist, or Bayes the clergyman—Graunt was a
seller of common notions: buttons, thread, needles, and other small
items used in a household. But Graunt wasn't just a button salesman;
he was a wealthy button salesman, and his wealth afforded him the
leisure to pursue interests having nothing to do with implements for
holding garments together. It also enabled him to befriend some of
the greatest intellectuals of his day, including Petty.

One inference Graunt gleaned from the mortality bills concerned the number of people who starved to death. In 1665 that
number was reported to be 45, only about double the number who
died from execution. In contrast, 4,808 were reported to have died
from consumption, 1,929 from "spotted fever and purples," 2,614
from "teeth and worms," and 68,596 from the plague. Why, when
London was "teeming with beggars," did so few starve? Graunt concluded that the populace must be feeding the hungry. And so he proposed instead that the state provide the food, thereby costing society
nothing while ridding seventeenth-century London streets of their
equivalent of panhandlers and squeegee men. Graunt also weighed
in on the two leading theories of how the plague is spread. One theory held that the illness was transmitted by foul air; the other, that it
was transmitted from person to person. Graunt looked at the week-to-
week records of deaths and concluded that the fluctuations in the
data were too great to be random, as he expected they would be if the
person-to-person theory were correct. On the other hand, since

weather varies erratically week by week, he considered the fluctuating data to be consistent with the foul-air theory. As it turned out, London was not ready for soup kitchens, and Londoners would have fared better if they had avoided ugly rats rather than foul air, but Graunt's great discoveries lay not in his conclusions. They lay instead in his realization that statistics can provide insights into the system from which the statistics are derived.

Petty's work is sometimes considered a harbinger of classical economics.[8] Believing that the strength of the state depends on, and is reflected by, the number and character of its subjects, Petty employed statistical reasoning to analyze national issues. Typically his analyses were made from the point of view of the sovereign and treated members of society as objects to be manipulated at will. Regarding the plague, he pointed out that money should be spent on prevention because, in saving lives, the realm would preserve part of the considerable investment society made in raising men and women to maturity and therefore would reap a higher return than it would on the most lucrative of alternative investments. Regarding the Irish, Petty was not as charitable. He concluded, for example, that the economic value of an English life was greater than that of an Irish one, so the wealth of the kingdom would be increased if all Irishmen except a few cowherds were forcibly relocated to England. As it happened, Petty owed his own wealth to those same Irish: as a doctor with the invading British army in the 1650s, he had been given the task of assessing the spoils and assessed that he could get away with grabbing a good share for himself, which he did.[9]

If, as Petty believed, the size and growth of a population reflect the quality of its government, then the lack of a good method for measuring the size of a population made the assessment of its government difficult. Graunt's most famous calculations addressed that issue — in particular the population of London. From the bills of mortality, Graunt knew the number of births. Since he had a rough idea of the fertility rate, he could infer how many women were of childbearing age. That datum allowed him to guess the total number of families and, using his own observations of the mean size of a London family, thereby estimate the city's population. He came up with 384,000—previously it was believed to be 2 million. Graunt also raised eyebrows by showing that much of the growth of the city was due to immigration from outlying areas, not to the slower method of procreation, and that despite the horrors of the plague, the decrease in population due to even the worst epidemic was always made up within two years. In addition, Graunt is generally credited with publishing the first "life table," a systematic arrangement of life-expectancy data that today is widely employed by organizations—from life insurance companies to the World Health Organization—that are interested in knowing how long people live. A life table displays how many people, in a group of 100, can be expected to survive to any given age. To Graunt's data (the column in the table below labeled "London, 1662"), I've added columns exhibiting the same data for a few countries today.[10]

Age	London, 1662	Afghanistan	Mozambique	China	Brazil	U.K.	Germany	U.S.	France	Japan
0	100	100	100	100	100	100	100	100	100	100
6	74	85	97	97	99	100	99	100	100	100
16	40	71	82	96	96	99	99	99	99	100
26	25	67	79	96	95	99	99	98	99	99
36	16	60	67	95	93	98	98	97	98	99
46	10	52	50	93	90	97	97	95	97	98
56	6	43	39	88	84	94	94	92	93	95
66	3	31	29	78	72	87	87	83	86	89
76	1	16	17	55	51	69	71	66	72	77
86	—	4	5	21	23	37	40	38	46	52
96	—	0	0	2	3	8	8	9	11	17

Graunt's life table extended

In 1662, Graunt published his analyses in *Natural and Political Observations . . . upon the Bills of Mortality*. The book met with acclaim. A year later Graunt was elected to the Royal Society. Then, in 1666, the Great Fire of London, which burned down a large part of the city, destroyed his business. To add insult to injury, he was accused of helping to cause the destruction by giving instructions to halt the water supply just before the fire started. In truth he had no affiliation with the water company until after the fire. Still, after that episode, Graunt's name disappeared from the books of the Royal Society. Graunt died of jaundice a few years later.

Largely because of Graunt's work, in 1667 the French fell in line with the British and revised their legal code to enable surveys like the bills of mortality. Other European countries followed suit. By the nineteenth century, statisticians all over Europe were up to their elbows in government records such as census data—"an avalanche of numbers."[11] Graunt's legacy was to demonstrate that inferences about a population as a whole could be made by carefully examining a limited sample of data. But though Graunt and others made valiant efforts to learn from the data through the application of simple logic, most of the data's secrets awaited the development of the tools created by Gauss, Laplace, and others in the nineteenth and early twentieth centuries.

THE TERM *statistics* entered the English language from the German word *Statistik* through a 1770 translation of the book *Bielfield's Elementary Universal Education*, which stated that "the science that is called statistics teaches us what is the political arrangement of all the modern states in the known world."[12] By 1828 the subject had evolved such that Noah Webster's *American Dictionary* defined statistics as "a collection of facts respecting the state of society, the condition of the people in a nation or country, their health, longevity, domestic economy, arts, property and political strength, the state of their country, &c."[13] The field had embraced the methods of Laplace, who had sought to extend his mathematical analysis from planets and stars to issues of everyday life.

The normal distribution describes the manner in which many phenomena vary around a central value that represents their most probable outcome; in his *Essai philosophique sur les probabilités*, Laplace argued that this new mathematics could be employed to assess legal testimony, predict marriage rates, calculate insurance premiums. But by the final edition of that work, Laplace was in his sixties, and so it fell to a younger man to develop his ideas. That man was Adolphe Quételet, born in Ghent, Flanders, on February 22, 1796.[14]

Quételet did not enter his studies spurred by a keen interest in the workings of society. His dissertation, which in 1819 earned him the first doctorate in science awarded by the new university in Ghent, was on the theory of conic sections, a topic in geometry. His interest then turned to astronomy, and around 1820 he became active in a movement to found a new observatory in Brussels, where he had taken a position. An ambitious man, Quételet apparently saw the observatory as a step toward establishing a scientific empire. It was an audacious move, not least because he knew relatively little about astronomy and virtually nothing about running an observatory. But he must have been persuasive, because not only did his observatory receive funding, but he personally received a grant to travel to Paris for several months to remedy the deficiencies in his knowledge. It proved a sound investment, for Quételet's Royal Observatory of Belgium is still in existence today.

In Paris, Quételet was affected in his own way by the disorder of life, and it pulled him in a completely different direction. His romance with statistics began when he made the acquaintance of several great French mathematicians, including Laplace and Joseph Fourier, and studied statistics and probability with Fourier. In the end, though he learned how to run an observatory, he fell in love with a different pursuit, the idea of applying the mathematical tools of astronomy to social data.

When Quételet returned to Brussels, he began to collect and analyze demographic data, soon focusing on records of criminal activity that the French government began to publish in 1827. In *Sur l'homme et le développement de ses facultés*, a two-volume work he published in 1835, Quételet printed a table of annual murders reported in France from 1826 to 1831. The number of murders, he noted, was relatively constant, as was the proportion of murders committed each year with guns, swords, knives, canes, stones, instruments for cutting and stabbing, kicks and punches, strangulation, drowning, and fire.[15] Quételet also analyzed mortality according to age, geography, season, and profession, as well as in hospitals and prisons. He studied statistics on drunkenness, insanity, and crime. And he discovered statistical regularities describing suicide by hanging in Paris and the number of marriages between sixty-something women and twenty-something men in Belgium.

Statisticians had conducted such studies before, but Quételet did something more with the data: he went beyond examining the average to scrutinizing the manner in which the data strayed from its average. Wherever he looked, Quételet found the normal distribution: in the propensities to crime, marriage, and suicide and in the height of American Indians and the chest measurements of Scottish soldiers (he came upon a sample of 5,738 chest measurements in an old issue of the *Edinburgh Medical and Surgical Journal*). In the height of 100,000 young Frenchmen called up for the draft he also found meaning in a deviation from the normal distribution. In that data, when the number of conscripts was plotted against their height, the bell-shaped curve was distorted: too few prospects were just above five feet two and a compensating surplus was just below that height. Quételet argued that the difference—about 2,200 extra "short men"—was due to fraud or, you might say friendly fudging, as those below five feet two were excused from service.

Decades later the great French mathematician Jules-Henri Poincaré employed Quételet's method to nab a baker who was short-changing his customers. At first, Poincaré, who made a habit of picking up a loaf of bread each day, noticed after weighing his loaves that they averaged about 950 grams instead of the 1,000 grams advertised. He complained to the authorities and afterward received bigger loaves. Still he had a hunch that something about his bread wasn't kosher. And so with the patience only a famous—or at least tenured—scholar can afford, he carefully weighed his bread every day for the next year. Though his bread now averaged closer to 1,000 grams, if the baker had been honestly handing him random loaves, the number of loaves heavier and lighter than the mean should have—as I mentioned in chapter 7—diminished following the bell-shaped pattern of the error law. Instead, Poincaré found that there were too few light loaves and a surplus of heavy ones. He concluded that the baker had not ceased baking underweight loaves but instead was seeking to placate him by always giving him the largest loaf he had on hand. The police again visited the cheating baker, who was reportedly appropriately astonished and presumably agreed to change his ways.[16]

Quételet had stumbled on a useful discovery: the patterns of randomness are so reliable that in certain social data their violation can be taken as evidence of wrongdoing. Today such analyses are applied to reams of data too large to have been analyzed in Quételet's time. In recent years, in fact, such statistical sleuthing has become popular, creating a new field, called forensic economics, perhaps the most famous example of which is the statistical study suggesting that companies were backdating their stock option grants. The idea is simple: companies grant stock options—the right to buy shares later at the price of the stock on the date of the grant—as an incentive for executives to improve their firms' share prices. If the grants are backdated to times when the shares were especially low, the executives' profits will be correspondingly high. A clever idea, but when done in secret it violates securities laws. It also leaves a statistical fingerprint, which has led to the investigation of such practices at about a dozen major companies.[17] In a less publicized example, Justin Wolfers, an economist at the Wharton School, found evidence of fraud in the results of about 70,000 college basketball games.[18]

Wolfers discovered the anomaly by comparing Las Vegas bookmakers' point spreads to the games' actual outcomes. When one team is favored, the bookmakers offer point spreads in order to attract a roughly even number of bets on both competitors. For instance, suppose the basketball team at Caltech is considered better than the team at UCLA (for college basketball fans, yes, this was actually true in the 1950s). Rather than assigning lopsided odds, bookies could instead offer an even bet on the game but pay out on a Caltech bet only if their team beat UCLA by, say, 13 or more points.

Though such point spreads are set by the bookies, they are really fixed by the mass of bettors because the bookies adjust them to balance the demand. (Bookies make their money on fees and seek to have an equal amount of money bet on each side so that they can't lose, whatever the outcome.) To measure how well bettors assess two teams, economists use a number called the forecast error, which is the difference between the favored team's margin of victory and the point spread determined by the marketplace. It may come as no surprise that forecast error, being a type of error, is distributed according to the normal distribution. Wolfers found that its mean is 0, meaning that the point spreads don't tend to either overrate or underrate teams, and its standard deviation is 10.9 points, meaning that about two thirds of the time the point spread is within 10.9 points of the margin of victory. (In a study of professional football games, a similar result was found, with a mean of 0 and a standard deviation of 13.9 points.)[19]

When Wolfers examined the subset of games that involved heavy favorites, he found something astonishing: there were too few games in which the heavy favorite won by a little more than the point spread and an inexplicable surplus of games in which the favorite won by just less. This was again Quételet's anomaly. Wolfers's conclusion, like Quételet's and Poincaré's, was fraud. His analysis went like this: it is hard for even a top player to ensure that his team will beat a point spread, but if the team is a heavy favorite, a player, without endangering his team's chance of victory, can slack off enough to ensure that the team does *not* beat the spread. And so if unscrupulous bettors wanted to fix games without asking players to risk losing, the result would be just the distortions Wolfers found. Does Wolfers's work prove that in some percentage of college basketball games, players are taking bribes to shave points? No, but as Wolfers says, "You shouldn't have what's happening on the court reflecting what's happening in Las Vegas." And it is interesting to note that in a recent poll by the National Collegiate Athletic Association, 1.5 percent of players admitted knowing a teammate "who took money for playing poorly."[20]

QUÉTELET DID NOT PURSUE the forensic applications of his ideas. He had bigger plans: to employ the normal distribution in order to illuminate the nature of people and society. If you made 1,000 copies of a statue, he wrote, those copies would vary due to errors of measurement and workmanship, and that variation would be governed by the error law. If the variation in people's physical traits follows the same law, he reasoned, it must be because we, too, are imperfect replicas of a prototype. Quételet called that prototype *l'homme moyen*, the average man. He felt that a template existed for human behavior too. The manager of a large department store may not know whether the spacey new cashier will pocket that half-ounce bottle of Chanel Allure she was sniffing, but he can count on the prediction that in the retail business, inventory loss runs pretty steadily from year to year at about 1.6 percent and that consistently about 45 percent to 48 percent of it is due to employee theft.[21] Crime, Quételet wrote, is "like a budget that is paid with frightening regularity."[22]

Quételet recognized that *l'homme moyen* would be different for different cultures and that it could change with changing social conditions. In fact, it is the study of those changes and their causes that was Quételet's greatest ambition. "Man is born, grows up, and dies according to certain laws," he wrote, and those laws "have never been studied."[23] Newton became the father of modern physics by recognizing and formulating a set of universal laws. Modeling himself after Newton, Quételet desired to create a new "social physics" describing the laws of human behavior. In Quételet's analogy, just as an object, if undisturbed, continues in its state of motion, so the mass behavior of people, if social conditions remain unchanged, remains constant. And just as Newton described how physical forces deflect an object from its straight path, so Quételet sought laws of human behavior describing how social forces transform the characteristics of society.

For example, Quételet thought that vast inequalities of wealth and great fluctuations in prices were responsible for crime and social unrest and that a steady level of crime represented a state of equilibrium, which would change with changes in the underlying causes. A vivid example of such a change in social equilibrium occurred in the months after the attacks of September 11, 2001, when travelers, afraid to take airplanes, suddenly switched to cars. Their fear translated into about 1,000 more highway fatalities in that period than in the same period the year before—hidden casualties of the September 11 attack.[24]

But to believe that a social physics exists is one thing, and to define one is another. In a true science, Quételet realized, theories could be explored by placing people in a great number of experimental situations and measuring their behavior. Since that is not possible, he concluded that social science is more like astronomy than physics, with insights deduced from passive observation. And so, seeking to uncover the laws of social physics, he studied the temporal and cultural variation in *l'homme moyen*.

Quételet's ideas were well received, especially in France and Great Britain. One physiologist even collected urine from a railroad-station urinal frequented by people of many nationalities in order to determine the properties of the "average European urine."[25] In Britain, Quételet's most enthusiastic disciple was a wealthy chess player and historian named Henry Thomas Buckle, best known for an ambitious multivolume book called *History of Civilization in England.* Unfortunately, in 1861, when he was forty, Buckle caught typhus while traveling in Damascus. Offered the services of a local physician, he refused because the man was French, and so he died. Buckle hadn't finished his treatise. But he did complete the initial two volumes, the first of which presented history from a statistical point of view. It was based on the work of Quételet and was an instant success. Read throughout Europe, it was translated into French, German, and Russian. Darwin read it; Alfred Russel Wallace read it; Dostoyevsky read it twice.[26]

Despite the book's popularity, the verdict of history is that

Quételet's mathematics proved more sensible than his social physics. For one thing, not all that happens in society, especially in the financial realm, is governed by the normal distribution. For example, if film revenue were normally distributed, most films would earn near some average amount, and two-thirds of all film revenue would fall within a standard deviation of that number. But in the film business, 20 percent of the movies bring in 80 percent of the revenue. Such hit-driven businesses, though thoroughly unpredictable, follow a far different distribution, one for which the concepts of mean and standard deviation have no meaning because there is no "typical" performance, and megahit outliers, which in an ordinary business might occur only once every few centuries, happen every few years.[27]

More important than his ignoring other probability distributions, though, is Quételet's failure to make much progress in uncovering the laws and forces he sought. So in the end his direct impact on the social sciences proved modest, yet his legacy is both undeniable and far-reaching. It lies not in the social sciences but in the "hard" sciences, where his approach to understanding the order in large numbers of random events inspired many scholars and spawned revolutionary work that transformed the manner of thinking in both biology and physics.

IT WAS CHARLES DARWIN'S FIRST COUSIN who introduced statistical thinking to biology. A man of leisure, Francis Galton had entered Trinity College, Cambridge, in 1840.[28] He first studied medicine but then followed Darwin's advice and changed his field to mathematics. He was twenty-two when his father died and he inherited a substantial sum. Never needing to work for a living, he became an amateur scientist. His obsession was measurement. He measured the size of people's heads, noses, and limbs, the number of times people fidgeted while listening to lectures, and the degree of attractiveness of girls he passed on the street (London girls scored highest; Aberdeen, lowest). He measured the characteristics of people's fingerprints, an endeavor that led to the adoption of fingerprint identification by Scotland Yard in 1901. He even measured the life spans of sovereigns and clergymen, which, being similar to the life spans of people in other professions, led him to conclude that prayer brought no benefit.

In his 1869 book, *Hereditary Genius*, Galton wrote that the fraction of the population in any given range of heights must be nearly uniform over time and that the normal distribution governs height and every other physical feature: circumference of the head, size of the brain, weight of the gray matter, number of brain fibers, and so on. But Galton didn't stop there. He believed that human character, too, is determined by heredity and, like people's physical features, obeys in some manner the normal distribution. And so, according to Galton, men are not "of equal value, as social units, equally capable of voting, and the rest."[29] Instead, he asserted, about 250 out of every 1 million men inherit exceptional ability in some area and as a result become eminent in their field. (As, in his day, women did not generally work, he did not make a similar analysis of them.) Galton founded a new field of study based on those ideas, calling it eugenics, from the Greek words *eu* (good) and *genos* (birth). Over the years, eugenics has meant many different things to many different people. The term and some of his ideas were adopted by the Nazis, but there is no evidence that Galton would have approved of the Germans' murderous schemes. His hope, rather, was to find a way to improve the condition of humankind through selective breeding.

Much of chapter 9 is devoted to understanding the reasons Galton's simple cause-and-effect interpretation of success is so seductive. But we'll see in chapter 10 that because of the myriad of foreseeable and chance obstacles that must be overcome to complete a task of any complexity, the connection between ability and accomplishment is far less direct than anything that can possibly be explained by Galton's ideas. In fact, in recent years psychologists have found that the ability to persist in the face of obstacles is at least as important a factor in success as talent.[30] That's why experts often speak of the "ten-year rule," meaning that it takes at least a decade of hard work, practice, and striving to become highly successful in most endeavors.

It might seem daunting to think that effort and chance, as much as innate talent, are what counts. But I find it encouraging because, while our genetic makeup is out of our control, our degree of effort is up to us. And the effects of chance, too, can be controlled to the extent that by committing ourselves to repeated attempts, we can increase our odds of success.

Whatever the pros and cons of eugenics, Galton's studies of inheritance led him to discover two mathematical concepts that are central to modern statistics. One came to him in 1875, after he distributed packets of sweet pea pods to seven friends. Each friend received seeds of uniform size and weight and returned to Galton the seeds of the successive generations. On measuring them, Galton noticed that the median diameter of the offspring of large seeds was less than that of the parents, whereas the median diameter of the offspring of small seeds was greater than that of the parents. Later, employing data he obtained from a laboratory he had set up in London, he noticed the same effect in the height of human parents and children. He dubbed the phenomenon—that in linked measurements, if one measured quantity is far from its mean, the other will be closer to its mean—regression toward the mean.

Galton soon realized that processes that did not exhibit regression toward the mean would eventually go out of control. For example, suppose the sons of tall fathers would on average be as tall as their fathers. Since heights vary, some sons would be taller. Now imagine the next generation, and suppose the sons of the taller sons, grandsons of the original men, were also on average as tall as their fathers. Some of them, too, would have to be taller than their fathers. In this way, as generation followed generation, the tallest humans would be ever taller. Because of regression toward the mean, that does not happen. The same can be said of innate intelligence, artistic talent, or the ability to hit a golf ball. And so very tall parents should not expect their children to be as tall, very brilliant parents should not expect their children to be as brilliant, and the Picassos and Tiger Woodses of this world should not expect their children to match their accomplishments. On the other hand, very short parents can expect taller offspring, and those of us who are not brilliant or can't paint have reasonable hope that our deficiencies will be improved upon in the next generation.

At his laboratory, Galton attracted subjects through advertisements and then subjected them to a series of measurements of height, weight, even the dimensions of certain bones. His goal was to find a method for predicting the measurements of children based on those of their parents. One of Galton's plots showed parents' heights versus the heights of their offspring. If, say, those heights were always equal, the graph would be a neat line rising at 45 degrees. If that relationship held on average but individual data points varied, then the data would show some scatter above and below that line. Galton's graphs thus exhibited visually not just the general relationship between the heights of parent and offspring but also the degree to which the relationship holds. That was Galton's other major contribution to statistics: defining a mathematical index describing the consistency of such relationships. He called it the coefficient of correlation.

The coefficient of correlation is a number between −1 and 1; if it is near ±1, it indicates that two variables are linearly related; a coefficient of 0 means there is no relation. For example, if data revealed that by eating the latest McDonald's 1,000-calorie meal once a week, people gained 10 pounds a year and by eating it twice a week they gained 20 pounds, and so on, the correlation coefficient would be 1. If for some reason everyone were to instead *lose* those amounts of weight, the correlation coefficient would be −1. And if the weight gain and loss were all over the map and didn't depend on meal consumption, the coefficient would be 0. Today correlation coefficients are among the most widely employed concepts in statistics. They are used to assess such relationships as those between the number of cigarettes smoked and the incidence of cancer, the distance of stars from Earth and the speed with which they are moving away from our planet, and the scores students achieve on standardized tests and the income of the students' families.

Galton's work was significant not just for its direct importance but because it inspired much of the statistical work done in the decades that followed, in which the field of statistics grew rapidly and matured. One of the most important of these advances was made by Karl Pearson, a disciple of Galton's. Earlier in this chapter, I mentioned many types of data that are distributed according to the normal distribution. But with a finite set of data the fit is never perfect. In the early days of statistics, scientists sometimes determined whether data were normally distributed simply by graphing them and observing the shape of the resulting curve. But how do you quantify the accuracy of the fit? Pearson invented a method, called the chi-square test, by which you can determine whether a set of data actually conforms to the distribution you believe it conforms to. He demonstrated his test in Monte Carlo in July 1892, performing a kind of rigorous repeat of Jagger's work.[31] In Pearson's test, as in Jagger's, the numbers that came up on a roulette wheel did not follow the distribution they would have followed if the wheel had produced random results. In another test, Pearson examined how many 5s and 6s came up in 26,306 tosses of twelve dice. He found that the distribution was not one you'd see in a chance experiment with fair dice—that is, in an experiment in which the probability of a 5 or a 6 on one roll were 1 in 3, or 0.3333. But it was consistent if the probability of a 5 or a 6 were 0.3377—that is, if the dice were skewed. In the case of the roulette wheel the game may have been rigged, but the dice were probably biased owing to variations in manufacturing, which my friend Moshe emphasized are always present.

Today chi-square tests are widely employed. Suppose, for instance, that instead of testing dice, you wish to test three cereal boxes for their consumer appeal. If consumers have no preference, you would expect about 1 in 3 of those polled to vote for each box. As we've seen, the actual results will rarely be distributed so evenly. Employing the chi-square test, you can determine how likely it is that the winning box received more votes due to consumer preference rather than to chance. Similarly, suppose researchers at a pharmaceutical company perform an experiment in which they test two treatments used in preventing acute transplant rejection. They can use a chi-square test to determine whether there is a statistically significant difference between the results. Or suppose that before opening a new outlet, the CFO of a rental car company expects that 25 percent of the company's customers will request subcompact cars, 50 percent will want compacts, and 12.5 percent each will ask for cars in the midsize and "other" categories. When the data begin to come in, a chi-square test can help the CFO quickly decide whether his assumption was correct or the new site is atypical and the company would do well to alter the mix.

Through Galton, Quételet's work infused the biological sciences. But Quételet also helped spur a revolution in the physical sciences: both James Clerk Maxwell and Ludwig Boltzmann, two of the founders of statistical physics, drew inspiration from Quételet's theories. (Like Darwin and Dostoyevsky, they read of them in Buckle's book.) After all, if the chests of 5,738 Scottish soldiers distribute themselves nicely along the curve of the normal distribution and the average yearly mileage of 200 million drivers can vary by as little as 100 miles from year to year, it doesn't take an Einstein to guess that the 10 septillion or so molecules in a liter of gas might exhibit some interesting regularities. But actually it did take an Einstein to finally convince the scientific world of the need for that new approach to physics. Albert Einstein did it in 1905, the same year in which he published his first work on relativity. And though hardly known in popular culture, Einstein's 1905 paper on statistical physics proved equally revolutionary. In the scientific literature, in fact, it would become his most cited work.[32]

EINSTEIN'S 1905 WORK on statistical physics was aimed at explaining a phenomenon called Brownian motion. The process was named for Robert Brown, botanist, world expert in microscopy, and the person credited with writing the first clear description of the cell nucleus. Brown's goal in life, pursued with relentless energy, was to discover through his observations the source of the life force, a mysterious influence believed in his day to endow something with the property of being alive. In that quest, Brown was doomed to failure, but one day in June 1827, he thought he had succeeded.

Peering through his lens, Brown noted that the granules inside the pollen grains he was observing seemed to be moving.[33] Though a source of life, pollen is not itself a living being. Yet as long as Brown stared, the movement never ceased, as if the granules possessed some mysterious energy. This was not intentioned movement; it seemed, in fact, to be completely random. With great excitement, Brown concluded at first that he had bagged his quarry, for what could this energy be but the energy that powers life itself?

In a string of experiments he performed assiduously over the next month, Brown observed the same kind of movement when suspending in water, and sometimes in gin, as wide a variety of organic particles as he could get his hands on: decomposing fibers of veal, spider's web "blackened with London dust," even his own mucus. Then, in a deathblow to his wishful interpretation of the discovery, Brown also observed the motion when looking at inorganic particles—of asbestos, copper, bismuth, antimony, and manganese. He knew then that the movement he was observing was unrelated to the issue of life. The true cause of Brownian motion would prove to be the same force that compelled the regularities in human behavior that Quételet had noted—not a physical force but an apparent force arising from the patterns of randomness. Unfortunately, Brown did not live to see this explanation of the phenomenon he observed.

The groundwork for the understanding of Brownian motion was laid in the decades that followed Brown's work, by Boltzmann, Maxwell, and others. Inspired by Quételet, they created the new field of statistical physics, employing the mathematical edifice of probability and statistics to explain how the properties of fluids arise from the movement of the (then hypothetical) atoms that make them up. Their ideas did not catch on for several more decades, however. Some scientists had mathematical issues with the theory. Others objected because at the time no one had ever seen an atom and no one believed anyone ever would. But most physicists are practical, and so the most important roadblock to acceptance was that although the theory reproduced some laws that were known, it made few new predictions. And so matters stood until 1905, when long after Maxwell was dead and shortly before a despondent Boltzmann would commit suicide, Einstein employed the nascent theory to explain in great numerical detail the precise mechanism of Brownian motion.[34] The necessity of a statistical approach to physics would never again be in doubt, and the idea that matter is made of atoms and molecules would prove to be the basis of most modern technology and one of the most important ideas in the history of physics.

The random motion of molecules in a fluid can be viewed, as we'll note in chapter 10, as a metaphor for our own paths through life, and so it is worthwhile to take a little time to give Einstein's work a closer look. According to the atomic picture, the fundamental motion of water molecules is chaotic. The molecules fly first this way, then that, moving in a straight line only until deflected by an encounter with one of their sisters. As mentioned in the Prologue, this type of path—in which at various points the direction changes randomly—is often called a drunkard's walk, for reasons obvious to anyone who has ever enjoyed a few too many martinis (more sober mathematicians and scientists sometimes call it a random walk). If particles that float in a liquid are, as atomic theory predicts, constantly and randomly bombarded by the molecules of the liquid, one might expect them to jiggle this way and that owing to the collisions. But there are two problems with that picture of Brownian motion: first, the molecules are far too light to budge the visible floating particles; second, molecular collisions occur far more frequently than the observed jiggles. Part of Einstein's genius was to realize that those two problems cancel each other out: though the collisions occur very frequently, because the molecules are so light, those frequent isolated collisions have no visible effect. It is only when pure luck occasionally leads to a lopsided preponderance of hits from some particular direction—the molecular analogue of Roger Maris's record year in baseball—that a noticeable jiggle occurs. When Einstein did the math, he found that despite the chaos on the microscopic level, there was a predictable relationship between factors such as the size, number, and speed of the molecules and the observable frequency and magnitude of the jiggling. Einstein had, for the first time, connected new and measurable consequences to statistical physics. That might sound like a largely technical achievement, but on the contrary, it represented the triumph of a great principle: that much of the order we perceive in nature belies an invisible underlying disorder and hence can be understood only through the rules of randomness. As Einstein wrote, "It is a magnificent feeling to recognize the unity of a complex of phenomena which appear to be things quite apart from the direct visible truth."[35]

In Einstein's mathematical analysis the normal distribution again played a central role, reaching a new place of glory in the history of science. The drunkard's walk, too, became established as one of the most fundamental—and soon one of the most studied—processes in nature. For as scientists in all fields began to accept the statistical approach as legitimate, they recognized the thumbprints of the drunkard's walk in virtually all areas of study—in the foraging of mosquitoes through cleared African jungle, in the chemistry of nylon, in the formation of plastics, in the motion of free quantum particles, in the movement of stock prices, even in the evolution of intelligence over eons of time. We'll examine the effects of randomness on our own paths through life in chapter 10. But as we're about to see, though in random variation there are orderly patterns, patterns are not always meaningful. And as important as it is to recognize the meaning when it is there, it is equally important not to extract meaning when it is not there. Avoiding the illusion of meaning in random patterns is a difficult task. It is the subject of the following chapter.

CHAPTER 9

Illusions of Patterns and Patterns of Illusion

I N 1848 TWO TEENAGE GIRLS, Margaret and Kate Fox, heard unexplained noises, like knocking or the moving of furniture. Their house, it happened, had a reputation for being haunted. As the story goes,[1] Kate challenged the source of the noises to repeat the snap of her fingers and to rap out her age. It rose to both challenges. Over the next few days, with their mother's and some neighbors' assistance, the sisters worked out a code with which they could communicate with the rapper (no pun intended). They concluded that the rapping originated with the spirit of a peddler who had been murdered years earlier in the home they now occupied. With that, modern spiritualism—the belief that the dead can communicate with the living—was born. By the early 1850s a particular type of spiritual contact, called table rapping, and its cousins, table moving and table turning, had become the rage in the United States and Europe. The enterprise consisted of a group of individuals arranging themselves around a table, resting their hands upon it, and waiting. In table rapping, after some time passed, a rap would be heard. In table moving and table turning, after time passed, the table would begin to tilt or move about, sometimes dragging the sitters along with it. One pictures serious bearded men with jackets reaching their midthigh and

ardent women in hoop skirts, eyes wide in wonder as their hands followed the table this way or that.

Table moving became so popular that in the summer of 1853 scientists began to look into it. One group of physicians noted that during the silent sitting period a kind of unconscious consensus seemed
to form about the direction in which the table would move.[2] They
found that when they diverted the sitters' attention so that a common
expectation could not form, the table did not move. In another trial
they managed to create a condition in which half the sitters expected
the table to move to the left and half expected it to move to the right,
and again it did not move. They concluded that "the motion was due
to muscular action, mostly exercised unconsciously." But the defini-
tive investigation was performed by the physicist Michael Faraday,
one of the founders of electromagnetic theory, inventor of the electric motor, and one of the greatest experimental scientists in history.[3]
Faraday first discovered that the phenomenon would occur even with
just one subject sitting at the table. Then, enrolling subjects who
were both "very honorable" and accomplished table movers, he conducted a series of ingenious and intricate experiments proving that
the movement of the sitters' hands preceded that of the table. Further, he designed an indicator that alerted the subjects in real time
whenever that was occurring. He found that "as soon as the . . . [indi-
cator] is placed before the most earnest [subject] . . . the power [of
the illusion] is gone; and this only because the parties are made conscious of what they are really doing."[4]

Faraday concluded, as the doctors had, that the sitters were
unconsciously pulling and pushing the table. The movement probably began as random fidgeting. Then at some point the sitters perceived in the randomness a pattern. That pattern precipitated a
self-fulfilling expectation as the subjects' hands followed the imag-
ined leadership of the table. The value of his indicator, Faraday
wrote, was thus "the corrective power it possesses over the mind of
the table-turner."[5] Human perception, Faraday recognized, is not a
direct consequence of reality but rather an act of imagination.[6]

Perception requires imagination because the data people encounter in their lives are never complete and always equivocal. For example, most people consider that the greatest evidence of an event one can obtain is to see it with their own eyes, and in a court of law little is held in more esteem than eyewitness testimony. Yet if you asked to display for a court a video of the same quality as the unprocessed data captured on the retina of a human eye, the judge might wonder what you were trying to put over. For one thing, the view will have a blind spot where the optic nerve attaches to the retina. Moreover, the only part of our field of vision with good resolution is a narrow area of about 1 degree of visual angle around the retina's center, an area the width of our thumb as it looks when held at arm's length. Outside that region, resolution drops off sharply. To compensate, we constantly move our eyes to bring the sharper region to bear on different portions of the scene we wish to observe. And so the pattern of raw data sent to the brain is a shaky, badly pixilated picture with a hole in it. Fortunately the brain processes the data, combining the input from both eyes, filling in gaps on the assumption that the visual properties of neighboring locations are similar and interpolating.[7] The result—at least until age, injury, disease, or an excess of mai tais takes its toll—is a happy human being suffering from the compelling illusion that his or her vision is sharp and clear.

We also use our imagination and take shortcuts to fill gaps in patterns of nonvisual data. As with visual input, we draw conclusions and make judgments based on uncertain and incomplete information, and we conclude, when we are done analyzing the patterns, that our "picture" is clear and accurate. But is it?

Scientists have moved to protect themselves from identifying false patterns by developing methods of statistical analysis to decide whether a set of observations provides good support for a hypothesis or whether, on the contrary, the apparent support is probably due to chance. For example, when physicists seek to determine whether the data from a supercollider is significant, they don't eyeball their graphs, looking for bumps that rise above the noise; they apply mathematical techniques. One such technique, significance testing, was developed in the 1920s by R. A. Fisher, one of the greatest statisticians of the twentieth century (a man also known for his uncontrol-lable temper and for a feud with his fellow statistics pioneer Karl Pearson that was so bitter he continued to attack his nemesis long after Pearson's death, in 1936).

To illustrate Fisher's ideas, suppose that a student in a research study on extrasensory perception predicts the result of some coin tosses. If in our observations we find that she is almost always right, we might hypothesize that she is somehow skilled at it, for instance, through psychic powers. On the other hand, if she is right about half the time, the data support the hypothesis that she was just guessing. But what if the data fall somewhere in between or if there isn't much data? Where do we draw the line between accepting and reject-ing the competing hypotheses? This is what significance testing does: it is a formal procedure for calculating the probability of our hav-ing observed what we observed *if* the hypothesis we are testing is true. If the probability is low, we reject the hypothesis. If it is high, we accept it.

For example, suppose we are skeptics and hypothesize that the student cannot accurately predict the results of coin tosses. And sup-pose that in an experimental trial she predicts the coin tosses cor-rectly in a certain number of cases. Then the methods we analyzed in chapter 4 allow us to calculate the probability that she could have accomplished the predictions by chance alone. If she had guessed the coin-toss results correctly so often that, say, the probability of her being that successful by chance alone is only 3 percent, then we would reject the hypothesis that she was guessing. In the jargon of significance testing, we would say the significance level of our rejec-tion is 3 percent, meaning that the chances are at most 3 percent that by chance the data has led us astray. A 3 percent level of significance is fairly impressive, and so the media might report the feat as new evi-dence of the existence of psychic powers. Still, those of us who don't believe in psychic powers might remain skeptical.

This example illustrates an important point: even with data signif-icant at, say, the 3 percent level, if you test 100 nonpsychic people for psychic abilities—or 100 ineffective drugs for their effectiveness—

you ought to expect a few people to show up as psychic or a few ineffective drugs to show up as effective. That's one reason political polls or medical studies, especially small ones, sometimes contradict earlier polls or studies. Still, significance testing and other statistical methods serve scientists well, especially when they can conduct large-scale controlled studies. But in everyday life we don't conduct such studies, nor do we intuitively apply statistical analysis. Instead, we rely on gut instinct. When my Viking stove turned out to be a lemon and by chance an acquaintance told me she'd had the same experience, I started telling my friends to avoid the brand. When the flight attendants on several United Airlines flights seemed grumpier than those on other airlines I'd recently flown with, I started avoiding United's flights. Not a lot of data there, but my gut instinct identified patterns.

Sometimes those patterns are meaningful. Sometimes they are not. In either case, the fact that our perception of the patterns of life is both highly convincing and highly subjective has profound implications. It implies a kind of relativity, a situation in which, as Faraday found, reality is in the eye of the beholder. For example, in 2006 *The New England Journal of Medicine* published a $12.5 million study of patients with documented osteoarthritis of the knee. The study showed that a combination of the nutritional supplements glucosamine and chondroitin is no more effective in relieving arthritis pain than a placebo. Still, one eminent doctor had a hard time letting go of his feeling that the supplements were effective and ended his analysis of the study on a national radio program by reaffirming the possible benefit of the treatment, remarking that, "One of my wife's doctors has a cat and she says that this cat cannot get up in the morning without a little dose of glucosamine and chondroitin sulfate."[8]

When we look closely, we find that many of the assumptions of modern society are based, as table moving is, on shared illusions. Whereas chapter 8 is concerned with the surprising regularities exhibited by random events, in what follows, I shall approach the issue from the opposite direction and examine how events whose patterns appear to have a definite cause may actually be the product of chance.

. . .

IT IS HUMAN NATURE to look for patterns and to assign them meaning when we find them. Kahneman and Tversky analyzed many of the shortcuts we employ in assessing patterns in data and in making judgments in the face of uncertainty. They dubbed those shortcuts heuristics. In general, heuristics are useful, but just as our manner of processing optical information sometimes leads to optical illusions, so heuristics sometimes lead to systematic error. Kahneman and Tversky called such errors biases. We all use heuristics, and we all suffer from biases. But although optical illusions seldom have much relevance in our everyday world, cognitive biases play an important role in human decision making. And so in the late twentieth century a movement sprang up to study how randomness is perceived by the human mind. Researchers concluded that "people have a very poor conception of randomness; they do not recognize it when they see it and they cannot produce it when they try,"[9] and what's worse, we routinely misjudge the role of chance in our lives and make decisions that are demonstrably misaligned with our own best interests.[10]

Imagine a sequence of events. The events might be quarterly earnings or a string of good or bad dates set up through an Internet dating service. In each case the longer the sequence, or the more sequences you look at, the greater the probability that you'll find every pattern imaginable—purely by chance. As a result, a string of good or bad quarters, or dates, need not have any "cause" at all. The point was rather starkly illustrated by the mathematician George Spencer-Brown, who wrote that in a random series of $10^{1,000,007}$ zeroes and ones, you should expect at least 10 nonoverlapping subsequences of 1 million consecutive zeros.[11] Imagine the poor fellow who bumps into one of those strings when attempting to use the random numbers for some scientific purpose. His software generates 5 zeros in a row, then 10, then 20, 1,000, 10,000, 100,000, 500,000. Would he be wrong to send back the program and ask for a refund? And how would a scientist react upon flipping open a newly pur-

174

chased book of random digits only to find that all the digits are zeros? Spencer-Brown's point was that there is a difference between a process being random and the product of that process appearing to be random. Apple ran into that issue with the random shuffling method it initially employed in its iPod music players: true randomness sometimes produces repetition, but when users heard the same song or songs by the same artist played back-to-back, they believed the shuffling wasn't random. And so the company made the feature "less random to make it feel more random," said Apple founder Steve Jobs.[12]

One of the earliest speculations about the perception of random patterns came from the philosopher Hans Reichenbach, who remarked in 1934 that people untrained in probability would have difficulty recognizing a random series of events.[13] Consider the following printout, representing the results of a sequence of 200 tosses of a coin, with X representing tails and O representing heads: oooo xxxxooooxxxooooxxooxooooxxooxxoooxxxooooxooxoxooooooxooxooooo oxxooxxxoxxxoxoxxxxoooxxooxxoxoooxxxooxooxoxoxxoxooooxoxooooooxx xxooooxxooxoxxoooxoooxxoxooxxooooxooxxxxooooxxxoooxoooxxxxxxx ooxxxooxooxoooooxxxx. It is easy to find patterns in the data—for instance, the four Os followed by four Xs at the beginning and the run of six Xs toward the end. According to the mathematics of randomness, such runs are to be expected in 200 random tosses. Yet they surprise most people. As a result, when instead of representing coin tosses, strings of Xs and Os represent events that affect our lives, people seek meaningful explanations for the pattern. When a string of Xs represents down days on the stock market, people believe the experts who explain that the market is jittery. When a string of Os represents a run of accomplishments by your favorite sports star, announcers sound convincing when they drone on about the player's "streakiness." And when, as we saw earlier, the Xs or Os stood for strings of failed films made by Paramount and Columbia Pictures, everyone nodded as the industry rags proclaimed just who did and who did not have a finger on the pulse of the worldwide movie audience.

Academics and writers have devoted much effort to studying patterns of random success in the financial markets. There is much evidence, for instance, that the performance of stocks is random—or so close to being random that in the absence of insider information and in the presence of a cost to make trades or manage your portfolio, you can't profit from any deviations from randomness.[14] Nevertheless, Wall Street has a long tradition of guru analysts, and the average analyst's salary, at the end of the 1990s, was about $3 million.[15] How do those analysts do? According to a 1995 study, the eight to twelve most highly paid "Wall Street superstars" invited by *Barron's* to make market recommendations at its annual roundtable merely matched the average market return.[16] Studies in 1987 and 1997 found that stocks recommended by the prognosticators on the television show *Wall $treet Week* did much worse, lagging far behind the market.[17] And in a study of 153 newsletters, a researcher at the Harvard Institute of Economic Research found "no significant evidence of stock-picking ability."[18]

By chance alone, some analysts and mutual funds will always exhibit impressive patterns of success. And though many studies show that these past market successes are not good indicators of future success—that is, that the successes were largely just luck— most people feel that the recommendations of their stockbrokers or the expertise of those running mutual funds are worth paying for. Many people, even intelligent investors, therefore buy funds that charge exorbitant management fees. In fact, when a group of savvy students from the Wharton business school were given a hypothetical $10,000 and prospectuses describing four index funds, each composed in order to mirror the S&P 500, the students overwhelmingly failed to choose the funds with the lowest fees.[19] Since paying even an extra 1 percent per year in fees could, over the years, diminish your retirement fund by as much as one-third or even one-half, the savvy students didn't exhibit very savvy behavior.

Of course, as Spencer-Brown's example illustrates, if you look long enough, you're bound to find someone who, through sheer luck, really has made startlingly successful predictions. For those who prefer real-world examples to mathematical scenarios involving

$10^{1,000,007}$ random digits, consider the case of the columnist Leonard Koppett.[20] In 1978, Koppett revealed a system that he claimed could determine, by the end of January every year, whether the stock market would go up or down in that calendar year. His system had correctly predicted the market, he said, for the past eleven years.[21] Of course, stock-picking systems are easy to identify in hindsight; the true test is whether they will work in the future. Koppett's system passed that test too: judging the market by the Dow Jones Industrial Average, it worked for eleven straight years, from 1979 through 1989, got it wrong in 1990, and was correct again every year until 1998. But although Koppett's predictions were correct for a streak of eighteen out of nineteen years, I feel confident in asserting that his streak involved no skill whatsoever. Why? Because Leonard Koppett was a columnist for *Sporting News,* and his system was based on the results of the Super Bowl, the championship game of professional football. Whenever the team from the (original) National Football League won, the stock market, he predicted, would rise. Whenever the team from the (original) American Football League won, he predicted, the market would go down. Given that information, few people would argue that Koppett was anything but lucky. Yet had he had different credentials—and not revealed his method—he could have been hailed as the most clever analyst since Charles H. Dow.

As a counterpoint to Koppett's story, consider now the story of a fellow who does have credentials, a fellow named Bill Miller. For years, Miller maintained a winning streak that, unlike Koppett's, was compared to Joe DiMaggio's fifty-six-game hitting streak and the seventy-four consecutive victories by the *Jeopardy!* quiz-show champ Ken Jennings. But in at least one respect these comparisons were not very apt: Miller's streak earned him each year more than those other gentlemen's streaks had earned them in their lifetimes. For Bill Miller was the sole portfolio manager of Legg Mason Value Trust Fund, and in each year of his fifteen-year streak his fund beat the portfolio of equity securities that constitute the Standard & Poor's 500.

For his accomplishments, Miller was heralded "the Greatest Money Manager of the 1990s" by *Money* magazine, "Fund Manager of the Decade" by Morningstar, and one of the top thirty most influential people in investing in 2001, 2003, 2004, 2005, and 2006 by SmartMoney.[22] In the fourteenth year of Miller's streak, one analyst was quoted on the CNNMoney Web site as putting the odds of a fourteen-year streak by chance alone at 372,529 to 1 (more on that later).[23]

Academics call the mistaken impression that a random streak is due to extraordinary performance the hot-hand fallacy. Much of the work on the hot-hand fallacy has been done in the context of sports because in sports, performance is easy to define and measure. Also, the rules of the game are clear and definite, data are plentiful and public, and situations of interest are replicated repeatedly. Not to mention that the subject gives academics a way to attend games and pretend they are working.

Interest in the hot-hand fallacy began around 1985, in particular with a paper by Tversky and his co-workers that was published in the journal Cognitive Psychology.[24] In that paper, "The Hot Hand in Basketball: On the Misperception of Random Sequences," Tversky and his colleagues investigated reams of basketball statistics. The players' talent varied, of course. Some made half their shots, some more, some less. Each player also had occasional hot and cold streaks. The paper's authors asked the question, how do the number and length of the streaks compare with what you would observe if the result of each shot were determined by a random process? That is, how would things have turned out if rather than shooting baskets, the players had tossed coins weighted to reflect their observed shooting percentages? The researchers found that despite the streaks, the floor shots of the Philadelphia 76ers, the free throws of the Boston Celtics, and the experimentally controlled floor shots of the Cornell University men's and women's varsity basketball teams exhibited no evidence of nonrandom behavior.

In particular, one direct indicator of "streakiness" is the conditional probability of success (that is, making a basket) *if* on the prior attempt the player had achieved success. For a streaky player, the chance of a success on the heels of a prior success should be higher than his or her overall chance of success. But the authors found that for each player a success following a success was just as likely as a success following a failure (that is, a missed basket).

A few years after Tversky's paper appeared, the Nobel Prize–winning physicist E. M. Purcell decided to investigate the nature of streaks in the sport of baseball.[25] As I mentioned in chapter 1, he found, in the words of his Harvard colleague Stephen Jay Gould, that except for Joe DiMaggio's fifty-six-game hitting streak, "nothing ever happened in baseball above and beyond the frequency predicted by coin-tossing models." Not even the twenty-one-game losing streak experienced at the start of the 1988 season by Major League Baseball's Baltimore Orioles. Bad players and teams have longer and more frequent streaks of failure than great players and great teams, and great players and great teams have longer and more frequent streaks of success than lesser players and lesser teams. But that is because their average failure or success rate is higher, and the higher the average rate, the longer and more frequent are the streaks that randomness will produce. To understand these events, you need only to understand the tossing of coins.

What about Bill Miller's streak? That a streak like Miller's could result from a random process may seem less shocking in light of a few other statistics. For instance, in 2004 Miller's fund gained just under 12 percent while the average stock in the S&P gained more than 15 percent.[26] It might sound like the S&P trounced Miller that year, but actually he counted 2004 in his "win" column. That is because the S&P 500 is not the simple average of the prices of the stocks it comprises; it is a weighted average in which stocks exert influence proportional to each company's capitalization. Miller's fund did worse than the simple average of S&P stocks but better than that weighted average. Actually, there were more than thirty twelve-month periods during his streak in which he lost to the weighted average, but they weren't calendar years, and the streak was based on the intervals from January 1 to December 31.[27] So the streak in a sense was an artificial one to start with, one that by chance was defined in a manner that worked for Miller.

But how can we reconcile these revelations with those 372,529-to-1 odds against him? In discussing Miller's streak in 2003, writers for *The Consilient Observer* newsletter (published by Credit Suisse–First Boston) said that "no other fund has ever outperformed the market for a dozen consecutive years in the last 40 years." They raised the question of the probability of a fund's accomplishing that by chance and went on to give three estimates of that probability (the year being 2003, they referred to the chances of a fund's beating the market for only twelve consecutive years): 1 in 4,096, 1 in 477,000, and 1 in 2.2 billion.[28] To paraphrase Einstein, if their estimates were correct, they would have needed only one. What were the actual chances? Roughly 3 out of 4, or 75 percent. That's quite a discrepancy, so I'd better explain.

Those who quoted the low odds were right in one sense: if you had singled out Bill Miller *in particular* at the start of 1991 *in particular* and calculated the odds that by pure chance *the specific person* you selected would beat the market *for precisely the next fifteen years*, then those odds would indeed have been astronomically low. You would have had the same odds against you if you had flipped a coin once a year for fifteen years with the goal of having it land heads up each time. But as in the Roger Maris home run analysis, those are not the relevant odds because there are thousands of mutual fund managers (over 6,000 currently), and there were many fifteen-year periods in which the feat could have been accomplished. So the relevant question is, if thousands of people are tossing coins once a year and have been doing so for decades, what are the chances that one of them, for some period of fifteen years or longer, will toss all heads? That probability is far, far higher than the odds of simply tossing fifteen heads in a row.

To make this explanation concrete, suppose 1,000 fund managers—certainly an underestimate—had each tossed a coin once a year starting in 1991 (the year Miller began his streak). After the first year about half of them would have tossed heads; after two years about one-quarter of them would have tossed two heads; after the third year one-eighth of them would have tossed three heads; and so on. By then some who had tossed tails would have started to drop out of the game, but that wouldn't affect the analysis because they had already failed. The chances that, after fifteen years, a *particular coin tosser* would have tossed all heads are then 1 in 32,768. But the chances that *someone among the 1,000* who had started tossing coins in 1991 would have tossed all heads are much higher, about 3 percent. Finally, there is no reason to consider only those who started tossing coins in 1991 — the fund managers could have started in 1990 or 1970 or any other year in the era of modern mutual funds. Since the writers for *The Consilient Observer* used forty years in their discussion, I calculated the odds that by chance *some manager* in the last four decades would beat the market each year for *some period of fifteen years or longer*. That latitude increased the odds again, to the probability I quoted earlier, almost 3 out of 4. So rather than being surprised by Miller's streak, I would say that if no one had achieved a streak like Miller's, you could have legitimately complained that all those highly paid managers were performing worse than they would have by blind chance!

I've cited some examples of the hot-hand fallacy in the context of sports and the financial world. But in all aspects of our lives we encounter streaks and other peculiar patterns of success and failure. Sometimes success predominates, sometimes failure. Either way it is important in our own lives to take the long view and understand that streaks and other patterns that don't appear random can indeed happen by pure chance. It is also important, when assessing others, to recognize that among a large group of people it would be very odd if one of them *didn't* experience a long streak of successes or failures.

No one credited Leonard Koppett for his lopsided successes, and no one would credit a coin tosser. Many people did credit Bill Miller. In his case, though the type of analysis I performed seems to have escaped many of the observers quoted in the media, it is no news to those who study Wall Street from the academic perspective. For example, the Nobel Prize–winning economist Merton Miller (no relation to Bill) wrote, "If there are 10,000 people looking at the stocks and trying to pick winners, one in 10,000 is going to score, by chance alone, and that's all that's going on. It's a game, it's a chance operation, and people think they are doing something purposeful but they're really not."[29] We must all draw our own conclusions depending on the circumstances, but with an understanding of how randomness operates, at least our conclusions need not be naive.

IN THE PRECEDING I've discussed how we can be fooled by the patterns in random sequences that develop over time. But random patterns in space can be just as misleading. Scientists know that one of the clearest ways to reveal the meaning of data is to display them in some sort of picture or graph. When we see data exhibited in this manner, meaningful relationships that we would likely have missed are often made obvious. The cost is that we also sometimes perceive patterns that in reality have no meaning. Our minds are made that way—to assimilate data, fill in gaps, and look for patterns. For example, look at the following arrangement of grayish squares in the figure below.

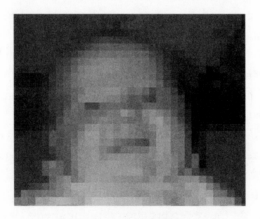

Photo from Frank H. Durgin,
"The Tinkerbell Effect,"
Journal of Consciousness Studies 9,
nos. 5–6 (May to June 2002)

The image does not literally look like a human being. Yet you can make enough sense of the pattern that if you saw in person the baby pictured, you would probably recognize it. And if you hold this book at arm's length and squint, you might not even perceive the imperfections in the image. Now look at this pattern of Xs and Os:

OOOOXXXX**OOO**XXX**OOOO**XX**OO**X**OOO**XXX**OO**XX**OOO**XXXX
OOOX**OO**X**O**X**OOOOO**X**OO**X**OOOOO**XX**OO**XXX**O**XX**O**X**O**XXXX
OOOXX**OO**XX**O**X**OO**XX**OO**X**OO**X**O**X**O**XX**O**X**OOO**X**O**X**OOOO**X
XXX**OOO**XX**OO**X**O**XX**OOO**X**OOO**XX**O**X**OO**XX**OOOO**X**OO**XXXX
OOOOXXX**OOO**X**OOO**XXXXXX**OO**XXX**OO**X**OO**X**OOOOO**XXXX

Here we see rectangular clusters, especially in the corners. I have put them in boldface. If the Xs and Os represented events of interest, we might be tempted to wonder if those clusters signified something. But any meaning we assigned them would be misconceived because these data are identical to the earlier set of 200 random Xs and Os, except for the geometric 5-by-40 arrangement and the choice of which letters to put in boldface.

This very issue drew much attention toward the end of World War II, when V2 rockets started raining down on London. The rockets were terrifying, traveling at over five times the speed of sound, so that one heard them approach only after they had hit. Newspapers soon published maps of the impact sites, which seemed to reveal not random patterns but purposeful clusters. To some observers the clusters indicated a precision in the control of the rockets' flight path that, given the distance the rockets had to travel, suggested that German technology was much more advanced than anyone had dreamed possible. Civilians speculated that the areas spared were home to German spies. Military leaders worried that the Germans could target crucial military sites, with devastating consequences.

In 1946 a mathematical analysis of the bombing data was published in the *Journal of the Institute of Actuaries*. Its author, R. D. Clarke, divided the area of interest into 576 parcels half a kilometer on each side. Of these, 229 parcels sustained no hits while, despite their minuscule size, 8 parcels had four or five hits. Still, Clarke's analysis showed that, like the coin-toss data above, the overall pattern was consistent with a random distribution.[30]

Similar issues arise frequently in reports of cancer clusters. If you divide any city or county into parcels and randomly distribute incidents of cancer, some parcels will receive less than average and some more. In fact, according to Raymond Richard Neutra, chief of the Division of Environmental and Occupational Disease Control in California's Department of Health, given a typical cancer registry— a database on local rates for dozens of different cancers—for California's 5,000 census tracts, you could expect to find 2,750 with statistically significant but random elevations of some form of cancer.[31] And if you look at a large enough number of such parcels, you'll find some regions in which cancer occurred at many times the normal rate.

The picture looks even worse if you draw the parcel boundaries *after* the cancers are distributed. What you get then is called the sharpshooter effect, after the apocryphal fellow who excels in his aim because he shoots at blank paper and draws the target afterward. Unfortunately that is how it usually happens in practice: first some citizens notice neighbors with cancer; then they define the boundaries of the area at issue. Thanks to the availability of data on the Internet, America these days is being scoured for such clusters. Not surprisingly, they are being found. Yet the development of cancer requires successive mutations. That means very long exposure and/or highly concentrated carcinogens. For such clusters of cancer to develop from environmental causes and show themselves in concert and before the victims have moved away from the affected area is quite a long shot. According to Neutra, to produce the kind of cancer clusters epidemiologists are typically called on to investigate, a population would have to be exposed to concentrations of carcinogens that are usually credible only in patients undergoing chemotherapy or in some work settings—far greater concentrations than people receive in contaminated neighborhoods and schools. Nevertheless, people resist accepting the explanation that the clusters are random fluctuations, and so each year state departments of health receive thousands of residential cancer-cluster reports, which result in the publication of hundreds of exhaustive analyses, none of which has convincingly identified an underlying environmental cause. Says Alan Bender, an epidemiologist with the Minnesota Department of Health, those studies "are an absolute, total, and complete waste of taxpayer dollars."[32]

So far in this chapter we have considered some of the ways in which random patterns can fool us. But psychologists have not contented themselves to merely study and categorize such misperceptions. They have also studied the reasons we fall prey to them. Let's now turn our attention to some of those factors.

PEOPLE LIKE TO EXERCISE CONTROL over their environment, which is why many of the same people who drive a car after consuming half a bottle of scotch will freak out if the airplane they are on experiences minor turbulence. Our desire to control events is not without purpose, for a sense of personal control is integral to our self-concept and sense of self-esteem. In fact, one of the most beneficial things we can do for ourselves is to look for ways to exercise control over our lives — or at least to look for ways that help us feel that we do. The psychologist Bruno Bettelheim observed, for instance, that survival in Nazi concentration camps "depended on one's ability to arrange to preserve some areas of independent action, to keep control of some important aspects of one's life despite an environment that seemed overwhelming."[33] Later studies showed that a prior sense of helplessness and lack of control is linked to both stress and the onset of disease. In one study wild rats were suddenly deprived of all control over their environment. They soon stopped struggling to survive and died.[34] In another study, in a group of subjects who were told they were going to take a battery of important tests, even the pointless power to control the order of those tests was found to reduce anxiety levels.[35]

One of the pioneers in the psychology of control is the psychologist and amateur painter Ellen Langer, now a professor at Harvard. Years ago, when she was at Yale, Langer and a collaborator studied the effect of the feeling of control on elderly nursing home patients.[36] One group was told they could decide how their rooms would be arranged and were allowed to choose a plant to care for. Another group had their rooms set up for them and a plant chosen and tended to for them. Within weeks the group that exercised control over their environment achieved higher scores on a predesigned measure of well-being. Disturbingly, eighteen months later a follow-up study shocked researchers: the group that was not given control experienced a death rate of 30 percent, whereas the group that was given control experienced a death rate of only 15 percent.[37]

Why is the human need to be in control relevant to a discussion of random patterns? Because if events are random, we are *not* in control, and if we are in control of events, they are *not* random. There is therefore a fundamental clash between our need to feel we are in control and our ability to recognize randomness. That clash is one of the principal reasons we misinterpret random events. In fact, inducing people to mistake luck for skill, or pointless actions for control, is one of the easiest enterprises a research psychologist can engage in. Ask people to control flashing lights by pressing a dummy button, and they will believe they are succeeding even though the lights are flashing at random.[38] Show people a circle of lights that flash at random and tell them that by concentrating they can cause the flashing to move in a clockwise direction, and they will astonish themselves with their ability to make it happen. Or have two groups simultaneously compete in a similar enterprise—one strives for clockwise motion along the circle, and the other attempts to make the lights travel counterclockwise—and the two groups will simultaneously perceive the lights traveling around the circle in the direction of their intention.[39]

Langer showed again and again how the need to feel in control interferes with the accurate perception of random events. In one of her studies, participants were found to be more confident of success when competing against a nervous, awkward rival than when competing against a confident one even though the card game in which they competed, and hence their probability of succeeding, was determined purely by chance.[40] In another study she asked a group of bright and well-educated Yale undergraduates to predict the results of thirty random coin tosses.[41] The experimenters secretly manipulated the outcomes so that each student was correct exactly half the time. They also arranged for some of the students to have early streaks of success. After the coin tosses the researchers quizzed the students in order to learn how they assessed their guessing ability. Many answered as if guessing a coin toss were a skill they could cultivate. One quarter reported that their performance would be hampered by distraction. Forty percent felt that their performance would improve with practice. And when asked directly to rate their ability at predicting the tosses, the students who achieved the early streaks of success rated themselves better at the task than did the others even though the number of successes was the same for all the subjects.

In another clever experiment, Langer set up a lottery in which each volunteer received a sports trading card with a player's picture on it.[42] A card identical to one of the distributed cards was placed in a bag with the understanding that the participant whose card it matched would be declared the winner. The players were divided into two groups. Those in one group had been allowed to choose their card; those in the other had been handed a card at random. Before the drawing each participant was given the opportunity to sell his or her card. Obviously, whether participants chose their cards or were handed them had no effect on their chances of winning. Yet those who had chosen their own cards demanded more than four times as much money for them as those selling the randomly assigned cards.

The subjects in Langer's experiments "knew," at least intellectually, that the enterprises in which they were engaging were random. When questioned, for example, none of the participants in the trading-card lottery said they believed that being allowed to choose their card had influenced their probability of winning. Yet they had *behaved* as if it had. Or as Langer wrote, "While people may pay lip service to the concept of chance, they behave as though chance events are subject to control."[43]

In real life the role of randomness is far less obvious than it was in Langer's experiments, and we are much more invested in the outcomes and our ability to influence them. And so in real life it is even more difficult to resist the illusion of control.

One manifestation of that illusion occurs when an organization experiences a period of improvement or failure and then readily attributes it not to the myriad of circumstances constituting the state of the organization as a whole and to luck but to the person at the top. That's especially obvious in sports, where, as I mentioned in the Prologue, if the players have a bad year or two, it is the coach who gets fired. In major corporations, in which operations are large and complex and to a great extent affected by unpredictable market forces, the causal connection between brilliance at the top and company performance is even less direct and the efficacy of reactionary firings is no greater than it is in sports. Researchers at Columbia University and Harvard, for example, recently studied a large number of corporations whose bylaws made them vulnerable to shareholders' demands that they respond to rough periods by changing management.[44] They found that in the three years after the firing there was no improvement, on average, in operating performance (a measure of earnings). No matter what the differences in ability among the CEOs, they were swamped by the effect of the uncontrollable elements of the system, just as the differences among musicians might become unapparent in a radio broadcast with sufficient noise and static. Yet in determining compensation, corporate boards of directors often behave as if the CEO is the *only* one who matters.

Research has shown that the illusion of control over chance events is enhanced in financial, sports, and especially, business situations when the outcome of a chance task is preceded by a period of strategizing (those endless meetings), when performance of the task requires active involvement (those long hours at the office), or when competition is present (this never happens, right?). The first step in battling the illusion of control is to be aware of it. But even then it is difficult, for, as we shall see in the following pages, once we think we see a pattern, we do not easily let go of our perception.

Suppose I tell you that I have made up a rule for the construction of a sequence of three numbers and that the sequence 2, 4, 6 satisfies my rule. Can you guess the rule? A single set of three numbers is not a lot to go on, so let's pretend that if you present me with other sequences of three numbers, I will tell you whether or not they satisfy my rule. Please take a moment to think up some three-number sequences to test—the advantage of reading a book over interacting in person is that in the book the author can display infinite patience.

Now that you have pondered your strategy, I can say that if you are like most people, the sequences you present will look something like 4, 6, 8 or 8, 10, 12 or 20, 24, 30. Yes, those sequences obey my rule. So what's the rule? Most people, after presenting a handful of such test cases, will grow confident and conclude that the rule is that the sequence must consist of increasing even numbers. But actually my rule was simply that the series must consist of increasing numbers. The sequence 1, 2, 3, for example, would have fit; there was no need for the numbers to be even. Would the sequences you thought of have revealed this?

When we are in the grasp of an illusion—or, for that matter, whenever we have a new idea—instead of searching for ways to prove our ideas wrong, we usually attempt to prove them correct. Psychologists call this the confirmation bias, and it presents a major impediment to our ability to break free from the misinterpretation of randomness. In the example above, most people immediately recognize that the sequence consists of increasing even numbers. Then, seeking to confirm their guess, they try out many more sequences of that type. But very few find the answer the fast way—through the attempt to falsify their idea by testing a sequence that includes an odd number.[45] As philosopher Francis Bacon put it in 1620, "the human understanding, once it has adopted an opinion, collects any instances that confirm it, and though the contrary instances may be more numerous and more weighty, it either does not notice them or else rejects them, in order that this opinion will remain unshaken."[46]

To make matters worse, not only do we preferentially seek evidence to confirm our preconceived notions, but we also interpret ambiguous evidence in favor of our ideas. This can be a big problem because data are often ambiguous, so by ignoring some patterns and emphasizing others, our clever brains can reinforce their beliefs even in the absence of convincing data. For instance, if we conclude, based on flimsy evidence, that a new neighbor is unfriendly, then any future actions that might be interpreted in that light stand out in our minds, and those that don't are easily forgotten. Or if we believe in a politician, then when she achieves good results, we credit her, and when she fails, we blame circumstances or the other party, either way reinforcing our initial ideas.

In one study that illustrated the effect rather vividly, researchers gathered a group of undergraduates, some of whom supported the death penalty and some of whom were against it.[47] The researchers then provided all the students with the same set of academic studies on the efficacy of capital punishment. Half the studies supported the idea that the death penalty has a deterrent effect; the other half contradicted that idea. The researchers also gave the subjects clues hinting at the weak points in each of the studies. Afterward the undergraduates were asked to rate the quality of the studies individually and whether and how strongly their attitudes about the death penalty were affected by their reading. The participants gave higher ratings to the studies that confirmed their initial point of view even when the studies on both sides had supposedly been carried out by the same method. And in the end, though everyone had read all the same studies, both those who initially supported the death penalty and those who initially opposed it reported that reading the studies had strengthened their beliefs. Rather than convincing anyone, the data polarized the group. Thus even random patterns can be interpreted as compelling evidence if they relate to our preconceived notions.

The confirmation bias has many unfortunate consequences in the real world. When a teacher initially believes that one student is smarter than another, he selectively focuses on evidence that tends to confirm the hypothesis.[48] When an employer interviews a prospective candidate, the employer typically forms a quick first impression and spends the rest of the interview seeking information that supports it.[49] When counselors in clinical settings are advised ahead of time that an interviewee is combative, they tend to conclude that he is even if the interviewee is no more combative than the average person.[50] And when people interpret the behavior of someone who is a member of a minority, they interpret it in the context of preconceived stereotypes.[51]

The human brain has evolved to be very efficient at pattern recognition, but as the confirmation bias shows, we are focused on finding and confirming patterns rather than minimizing our false conclusions. Yet we needn't be pessimists, for it is possible to overcome our prejudices. It is a start simply to realize that chance events, too, produce patterns. It is another great step if we learn to question our perceptions and our theories. Finally, we should learn to spend as much time looking for evidence that we are wrong as we spend searching for reasons we are correct.

Our journey through randomness is now almost at its end. We began with simple rules and went on to learn how they reflect themselves in complex systems. How great is the role of chance in that most important complex system of all—our personal destiny? That's a difficult question, one that has infused much of what we have considered thus far. And though I can't hope to answer it fully, I do hope to shed light on it. My conclusion is evident from the following chapter's title, which is the same as that of this book: "The Drunkard's Walk."

CHAPTER 10

The Drunkard's Walk

IN 1814, near the height of the great successes of Newtonian physics, Pierre-Simon de Laplace wrote:

> If an intelligence, at a given instant, knew all the forces that animate nature and the position of each constituent being; if, moreover, this intelligence were sufficiently great to submit these data to analysis, it could embrace in the same formula the movements of the greatest bodies in the universe and those of the smallest atoms: to this intelligence nothing would be uncertain, and the future, as the past, would be present to its eyes.[1]

Laplace was expressing a view called determinism: the idea that the state of the world at the present determines precisely the manner in which the future will unfold.

In everyday life, determinism implies a world in which our personal qualities and the properties of any given situation or environment lead directly and unequivocally to precise consequences. That is an orderly world, one in which everything can be foreseen, computed, predicted. But for Laplace's dream to hold true, several conditions must be met. First, the laws of nature must dictate a definite future, and we must know those laws. Second, we must have access to

192

data that completely describe the system of interest, allowing no unforeseen influences. Finally, we must have sufficient intelligence or computing power to be able to decide what, given the data about the present, the laws say the future will hold. In this book we've examined many concepts that aid our understanding of random phenomena. Along the way we've gained insight into a variety of specific life situations. Yet there remains the big picture, the question of how much randomness contributes to where we are in life and how well we can predict where we are going.

In the study of human affairs from the late Renaissance to the Victorian era, many scholars shared Laplace's belief in determinism. They felt as Galton did that our path in life is strictly determined by our personal qualities, or like Quételet they believed that the future of society is predictable. Often they were inspired by the success of Newtonian physics and believed that human behavior could be foretold as reliably as other phenomena in nature. It seemed reasonable to them that the future events of the everyday world should be as rigidly determined by the present state of affairs as are the orbits of the planets.

In the 1960s a meteorologist named Edward Lorenz sought to employ the newest technology of his day—a primitive computer—to carry out Laplace's program in the limited realm of the weather. That is, if Lorenz supplied his noisy machine with data on the atmospheric conditions of his idealized earth at some given time, it would employ the known laws of meteorology to calculate and print out rows of numbers representing the weather conditions at future times.

One day, Lorenz decided he wanted to extend a particular simulation further into the future. Instead of repeating the entire calculation, he decided to take a shortcut by beginning the calculation midway through. To accomplish that, he employed as initial conditions data printed out in the earlier simulation. He expected the computer to regenerate the remainder of the previous simulation and then carry it further. But instead he noticed something strange: the weather had evolved differently. Rather than duplicating the end of the previous simulation, the new one diverged wildly. He soon recognized why: in the computer's memory the data were stored to six decimal places, but in the printout they were quoted to only three. As a result, the data he had supplied were a tiny bit off. A number like 0.293416, for example, would have appeared simply as 0.293.

Scientists usually assume that if the initial conditions of a system are altered slightly, the evolution of that system, too, will be altered slightly. After all, the satellites that collect weather data can measure parameters to only two or three decimal places, and so they cannot even track a difference as tiny as that between 0.293416 and 0.293. But Lorenz found that such small differences led to massive changes in the result.[2] The phenomenon was dubbed the butterfly effect, based on the implication that atmospheric changes so small they could have been caused by a butterfly flapping its wings can have a large effect on subsequent global weather patterns. That notion might sound absurd—the equivalent of the extra cup of coffee you sip one morning leading to profound changes in your life. But actually that does happen—for instance, if the extra time you spent caused you to cross paths with your future wife at the train station or to miss being hit by a car that sped through a red light. In fact, Lorenz's story is itself an example of the butterfly effect, for if he hadn't taken the minor decision to extend his calculation employing the shortcut, he would not have discovered the butterfly effect, a discovery which sparked a whole new field of mathematics. When we look back in detail on the major events of our lives, it is not uncommon to be able to identify such seemingly inconsequential random events that led to big changes.

Determinism in human affairs fails to meet the requirements for predictability alluded to by Laplace for several reasons. First, as far as we know, society is not governed by definite and fundamental laws in the way physics is. Instead, people's behavior is not only unpredictable, but as Kahneman and Tversky repeatedly showed, also often irrational (in the sense that we act against our best interests). Second, even if we could uncover the laws of human affairs, as Quételet attempted to do, it is impossible to precisely know or control the circumstances of life. That is, like Lorenz, we cannot obtain the precise data necessary for making predictions. And third, human affairs are so complex that it is doubtful we could carry out the necessary calculations even if we understood the laws and possessed the data. As a result, determinism is a poor model for the human experience. Or as the Nobel laureate Max Born wrote, "Chance is a more fundamental conception than causality."[3]

In the scientific study of random processes the drunkard's walk is the archetype. In our lives it also provides an apt model, for like the granules of pollen floating in the Brownian fluid, we're continually nudged in this direction and then that one by random events. As a result, although statistical regularities can be found in social data, the future of particular individuals is impossible to predict, and for our particular achievements, our jobs, our friends, our finances, we all owe more to chance than many people realize. On the following pages, I shall argue, furthermore, that in all except the simplest real-life endeavors unforeseeable or unpredictable forces cannot be avoided, and moreover those random forces and our reactions to them account for much of what constitutes our particular path in life. I will begin my argument by exploring an apparent contradiction to that idea: if the future is really chaotic and unpredictable, why, after events have occurred, does it often seem as if we should have been able to foresee them?

IN THE FALL OF 1941, a few months before the Japanese attack on Pearl Harbor, an agent in Tokyo sent a spy in Honolulu an alarming request.[4] The request was intercepted and sent to the Office of Naval Intelligence. It wended its way through the bureaucracy, reaching Washington in decoded and translated form on October 9. The message requested the Japanese agent in Honolulu to divide Pearl Harbor into five areas and to make reports on ships in the harbor with reference to those areas. Of special interest were battleships, destroyers, and aircraft carriers, as well as information regarding the anchoring of more than one ship at a single dock. Some weeks later another curious incident occurred: U.S. monitors lost track of radio communications from all known carriers in the first and second Japanese fleets, losing with it all knowledge of their whereabouts. Then in early December the Combat Intelligence Unit of the Fourteenth Naval District in Hawaii reported that the Japanese had changed their call signs for the second time in a month. Call signs, such as WCBS or KNPR, are designations identifying the source of a radio transmission. In wartime they reveal the identity of a source not only to friend but also to foe, so they are periodically altered. The Japanese had a habit of changing them every six months or more. To change them twice in thirty days was considered a "step in preparing for active operations on a large scale." The change made identification of the whereabouts of Japanese carriers and submarines in the ensuing days difficult, further confusing the issue of the radio silence.

Two days later messages sent to Japanese diplomatic and consular posts at Hong Kong, Singapore, Batavia, Manila, Washington, and London were intercepted and decoded. They called for the diplomats to destroy most of their codes and ciphers immediately and to burn all other important confidential and secret documents. Around that time the FBI also intercepted a telephone call from a cook at the Japanese consulate in Hawaii to someone in Honolulu reporting in great excitement that the officials there were burning all major documents. The assistant head of the main unit of army intelligence, Lieutenant Colonel George W. Bicknell, brought one of the intercepted messages to his boss as he was preparing to go to dinner with the head of the army's Hawaiian Department. It was late afternoon on Saturday, December 6, the day before the attack. Bicknell's higher-up took five minutes to consider the message, then dismissed it and went to eat. With events so foreboding when considered in hindsight, why hadn't anyone privy to this information seen the attack coming?

In any complex string of events in which each event unfolds with some element of uncertainty, there is a fundamental asymmetry between past and future. This asymmetry has been the subject of scientific study ever since Boltzmann made his statistical analysis of the molecular processes responsible for the properties of fluids (see chapter 8). Imagine, for example, a dye molecule floating in a glass of water. The molecule will, like one of Brown's granules, follow a drunkard's walk. But even that aimless movement makes progress in some direction. If you wait three hours, for example, the molecule will typically have traveled about an inch from where it started. Suppose that at some point the molecule moves to a position of significance and so finally attracts our attention. As many did after Pearl Harbor, we might look for the reason why that unexpected event occurred. Now suppose we dig into the molecule's past. Suppose, in fact, we trace the record of all its collisions. We will indeed discover how first this bump from a water molecule and then that one propelled the dye molecule on its zigzag path from there to here. In hindsight, in other words, we can clearly explain why the past of the dye molecule developed as it did. But the water contains many other water molecules that *could have* been the ones that interacted with the dye molecule. To predict the dye molecule's path *beforehand* would have therefore required us to calculate the paths and mutual interactions of all those potentially important water molecules. That would have involved an almost unimaginable number of mathematical calculations, far greater in scope and difficulty than the list of collisions needed to understand the past. In other words, the movement of the dye molecule was virtually impossible to predict before the fact even though it was relatively easy to understand afterward.

That fundamental asymmetry is why in day-to-day life the past often seems obvious even when we could not have predicted it. It's why weather forecasters can tell you the reasons why three days ago the cold front moved like this and yesterday the warm front moved like that, causing it to rain on your romantic garden wedding, but the same forecasters are much less successful at knowing how the fronts will behave three days hence and at providing the warning you would have needed to get that big tent ready. Or consider a game of chess. Unlike card games, chess involves no explicit random element. And yet there is uncertainty because neither player knows for sure what his or her opponent will do next. If the players are expert, at most points in the game it may be possible to see a few moves into the future; if you look out any further, the uncertainty will compound, and no one will be able to say with any confidence exactly how the game will turn out. On the other hand, looking back, it is usually easy to say why each player made the moves he or she made. This again is a probabilistic process whose future is difficult to predict but whose past is easy to understand.

The same thing is true of the stock market. Consider, for instance, the performance of mutual funds. As I mentioned in chapter 9, it is common, when choosing a mutual fund, to look at past performance. Indeed, it is easy to find nice, orderly patterns when looking back. Here, for example, is a graph of the performance of 800 mutual fund managers over the five-year period, 1991–1995.

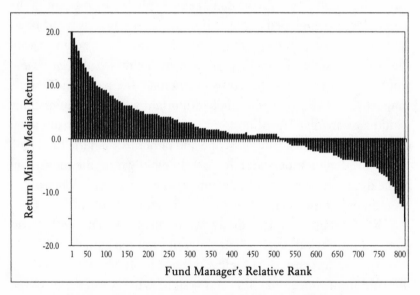

Performance versus ranking of the top mutual funds
in the five-year period 1991–1995.

On the vertical axis are plotted the funds' gains or losses relative to the average fund of the group. In other words, a return of 0 percent means the fund's performance was average for this five-year period. On the horizontal axis is plotted the managers' relative rank, from the number-1 performer to the number-800 performer. To look up the performance of, say, the 100th most successful mutual fund manager in the given five-year period, you find the point on the graph corresponding to the spot labeled 100 on the horizontal axis.

Any analyst, no doubt, could give a number of convincing reasons why the top managers represented here succeeded, why the bottom ones failed, and why the curve should take this shape. And whether or not we take the time to follow such analyses in detail, few are the investors who would choose a fund that has performed 10 percent below average in the past five years over a fund that has done 10 percent better than average. It is easy, looking at the past, to construct such nice graphs and neat explanations, but this logical picture of events is just an illusion of hindsight with little relevance for predicting future events. In the graph on page 200, for example, I compare how the same funds, still ranked in order of their performance in the *initial* five-year period, performed in the *next* five-year period. In other words, I maintain the ranking based on the period 1991–1995, but display the return the funds achieved in 1996–2000. If the past were a good indication of the future, the funds I considered in the period 1991–1995 would have had more or less the same relative performance in 1996–2000. That is, if the winners (at the left of the graph) continued to do better than the others, and the losers (at the right) worse, this graph should be nearly identical to the last. Instead, as we can see, the order of the past dissolves when extrapolated to the future, and the graph ends up looking like random noise.

People systematically fail to see the role of chance in the success of ventures and in the success of people like the equity-fund manager Bill Miller. And we unreasonably believe that the mistakes of the past must be consequences of ignorance or incompetence and could have been remedied by further study and improved insight. That's why, for example, in spring 2007, when the stock of Merrill Lynch was trading around $95 a share, its CEO E. Stanley O'Neal could be celebrated as the risk-taking genius responsible, and in the fall of 2007, after the credit market collapsed, derided as the risk-taking cowboy responsible—and promptly fired. We afford automatic re-

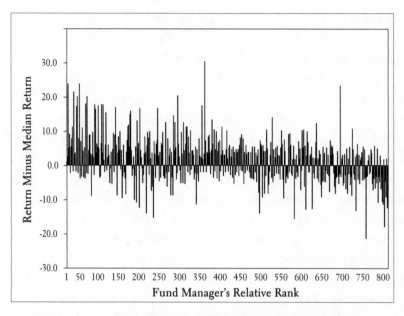

How the top funds in 1991–1995 performed in 1996–2000.

spect to superstar business moguls, politicians, and actors and to anyone flying around in a private jet, as if their accomplishments must reflect unique qualities not shared by those forced to eat commercial-airline food. And we place too much confidence in the overly precise predictions of people—political pundits, financial experts, business consultants—who claim a track record demonstrating expertise.

One large publishing company I'm familiar with went to great pains to develop one-year, three-year, and five-year plans for its educational software division. There were high-paid consultants, lengthy marketing meetings, late-night financial-analysis sessions, long off-site afternoon powwows. In the end, hunches were turned into formulas claiming the precision of several decimal places, and wild guesses were codified as likely outcomes. When in the first year certain products didn't sell as well as expected or others sold better than projected, reasons were found and the appropriate employees blamed or credited as if the initial expectations had been meaningful. The next year saw a series of unforeseen price wars started by two competitors. The year after that the market for educational software collapsed. As the uncertainty compounded, the three-year plan never had a chance to succeed. And the five-year plan, polished and precise as a diamond, was spared any comparison with performance, for by then virtually everyone in the division had moved on to greener pastures.

Historians, whose profession is to study the past, are as wary as scientists of the idea that events unfold in a manner that can be predicted. In fact, in the study of history the illusion of inevitability has such serious consequences that it is one of the few things that both conservative and socialist historians can agree on. The socialist historian Richard Henry Tawney, for example, put it like this: "Historians give an appearance of inevitability . . . by dragging into prominence the forces which have triumphed and thrusting into the background those which they have swallowed up."[5] And the historian Roberta Wohlstetter, who received the Presidential Medal of Freedom from Ronald Reagan, said it this way: "After the event, of course, a signal is always crystal clear; we can now see what disaster it was signaling. . . . But before the event it is obscure and pregnant with conflicting meanings."[6]

In some sense this idea is encapsulated in the cliché that hindsight is always 20/20, but people often behave as if the adage weren't true. In government, for example, a should-have-known-it blame game is played after every tragedy. In the case of Pearl Harbor (and the September 11 attacks) the events leading up to the attack, when we look back at them, certainly seem to point in an obvious direction. Yet as with the dye molecule, the weather, or the chess game, if you start well before the fact and trace events forward, the feeling of inevitability quickly dissolves. For one thing, in addition to the intelligence reports I quoted, there was a huge amount of useless intelligence, with each week bringing new reams of sometimes alarming or mysterious messages and transcripts that would later prove misleading or insignificant. And even if we focused on the reports that hindsight tells us were important, before the attack there existed for each of those reports a reasonable alternative explanation that did not point toward a surprise attack on Pearl Harbor. For example, the request to divide Pearl Harbor into five areas was similar in style to other requests that had gone to Japanese agents in Panama, Vancouver, San Francisco, and Portland, Oregon. The loss of radio contact was also not unheard of and had in the past often meant simply that the warships were in home waters and communicating via telegraphic landlines. Moreover, even if you believed a broadening of the war was impending, many signs pointed toward an attack elsewhere—in the Philippines, on the Thai peninsula, or on Guam, for example. There were not, to be sure, as many red herrings as water molecules encountered by a molecule of dye, but there were enough to obscure a clear vision of the future.

After Pearl Harbor seven committees of the U.S. Congress delved into the process of discovering why the military had missed all the "signs" of a coming attack. Army Chief of Staff General George Marshall, for one, drew heavy criticism for a May 1941 memo to President Roosevelt in which he wrote that "the Island of Oahu, due to its fortification, its garrison and its physical characteristic, is believed to be the strongest fortress in the world" and reassured the president that, in case of attack, enemy forces would be intercepted "within 200 miles of their objective . . . by all types of bombardment." General Marshall was no fool, but neither did he have a crystal ball. The study of randomness tells us that the crystal ball view of events is possible, unfortunately, only after they happen. And so we believe we know why a film did well, a candidate won an election, a storm hit, a stock went down, a soccer team lost, a new product failed, or a disease took a turn for the worse, but such expertise is empty in the sense that it is of little use in predicting when a film will do well, a candidate will win an election, a storm will hit, a stock will go down, a soccer team will lose, a new product will fail, or a disease will take a turn for the worse.

It is easy to concoct stories explaining the past or to become confident about dubious scenarios for the future. That there are traps in such endeavors doesn't mean we should not undertake them. But we can work to immunize ourselves against our errors of intuition. We can learn to view both explanations and prophecies with skepticism.

We can focus on the ability to react to events rather than relying on the ability to predict them, on qualities like flexibility, confidence, courage, and perseverance. And we can place more importance on our direct impressions of people than on their well-trumpeted past accomplishments. In these ways we can resist forming judgments in our automatic deterministic framework.

In MARCH 1979 another famously unanticipated chain of events occurred, this one at a nuclear power plant in Pennsylvania.[7] It resulted in a partial meltdown of the core, in which the nuclear reaction occurs, threatening to release into the environment an alarming dose of radiation. The mishap began when a cup or so of water emerged through a leaky seal from a water filter called a polisher. The leaked water entered a pneumatic system that drives some of the plant's instruments, tripping two valves. The tripped valves shut down the flow of cold water to the plant's steam generator—the system responsible for removing the heat generated by the nuclear reaction in the core. An emergency water pump then came on, but a valve in each of its two pipes had been left in a closed position after maintenance two days earlier. The pumps therefore were pumping water uselessly toward a dead end. Moreover, a pressure-relief valve also failed, as did a gauge in the control room that ought to have shown that the valve was not working.

Viewed separately, each of the failures was of a type considered both commonplace and acceptable. Polisher problems were not unusual at the plant, nor were they normally very serious; with hundreds of valves regularly being opened or closed in a nuclear power plant, leaving some valves in the wrong position was not considered rare or alarming; and the pressure-relief valve was known to be somewhat unreliable and had failed at times without major consequences in at least eleven other power plants. Yet strung together, these failures make the plant seem as if it had been run by the Keystone Kops. And so after the incident at Three Mile Island came many investigations and much laying of blame, as well as a very different consequence. That string of events spurred Yale sociologist Charles Perrow to create a new theory of accidents, in which is codified the central argument of this chapter: in complex systems (among which I count our lives) we should expect that minor factors we can usually ignore will by chance sometimes cause major incidents.[8]

In his theory Perrow recognized that modern systems are made up of thousands of parts, including fallible human decision makers, which interrelate in ways that are, like Laplace's atoms, impossible to track and anticipate individually. Yet one can bet on the fact that just as atoms executing a drunkard's walk will eventually get somewhere, so too will accidents eventually occur. Called normal accident theory, Perrow's doctrine describes how that happens—how accidents can occur without clear causes, without those glaring errors and incompetent villains sought by corporate or government commissions. But although normal accident theory is a theory of why, inevitably, things sometimes go wrong, it could also be flipped around to explain why, inevitably, they sometimes go right. For in a complex undertaking, no matter how many times we fail, if we keep trying, there is often a good chance we will eventually succeed. In fact, economists like W. Brian Arthur argue that a concurrence of minor factors can even lead companies with no particular edge to come to dominate their competitors. "In the real world," he wrote, "if several similar-sized firms entered a market together, small fortuitous events—unexpected orders, chance meetings with buyers, managerial whims—would help determine which ones received early sales and, over time, which came to dominate. Economic activity is . . . [determined] by individual transactions that are too small to foresee, and these small 'random' events could [ac]cumulate and become magnified by positive feedbacks over time."[9]

The same phenomenon has been noticed by researchers in sociology. One group, for example, studied the buying habits of consumers in what sociologists call the cultural industries—books, film, art, music. The conventional marketing wisdom in those fields is that success is achieved by anticipating consumer preference. In this view the most productive way for executives to spend their time is to study what it is about the likes of Stephen King, Madonna, or Bruce Willis that appeals to so many fans. They study the past and, as I've just argued, have no trouble extracting reasons for whatever success they are attempting to explain. They then try to replicate it.

That is the deterministic view of the marketplace, a view in which it is mainly the intrinsic qualities of the person or the product that governs success. But there is another way to look at it, a nondeterministic view. In this view there are many high-quality but unknown books, singers, actors, and what makes one or another come to stand out is largely a conspiracy of random and minor factors—that is, luck. In this view the traditional executives are just spinning their wheels.

Thanks to the Internet, this idea has been tested. The researchers who tested it focused on the music market, in which Internet sales are coming to dominate. For their study they recruited 14,341 participants who were asked to listen to, rate, and if they desired, download 48 songs by bands they had not heard of.[10] Some of the participants were also allowed to view data on the popularity of each song—that is, on how many fellow participants had downloaded it. These participants were divided into eight separate "worlds" and could see only the data on downloads of people in their own world. All the artists in all the worlds began with zero downloads, after which each world evolved independently. There was also a ninth group of participants, who were not shown any data. The researchers employed the popularity of the songs in this latter group of insulated listeners to define the "intrinsic quality" of each song—that is, its appeal in the absence of external influence.

If the deterministic view of the world were true, the same songs ought to have dominated in each of the eight worlds, and the popularity rankings in those worlds ought to have agreed with the intrinsic quality as determined by the isolated individuals. But the researchers found exactly the opposite: the popularity of individual songs varied widely among the different worlds, and different songs of similar intrinsic quality also varied widely in their popularity. For example, a song called "Lockdown" by a band called 52metro ranked twenty-six out of forty-eight in intrinsic quality but was the number-1 song in one world and the number-40 song in another. In this experiment, as one song or another by chance got an early edge in downloads, its seeming popularity influenced future shoppers. It's a phenomenon that is well-known in the movie industry: moviegoers will report liking a movie more when they hear beforehand how good it is. In this example, small chance influences created a snowball effect and made a huge difference in the future of the song. Again, it's the butterfly effect.

In our lives, too, we can see through the microscope of close scrutiny that many major events would have turned out differently were it not for the random confluence of minor factors, people we've met by chance, job opportunities that randomly came our way. For example, consider the actor who, for seven years starting in the late 1970s, lived in a fifth-floor walk-up on Forty-ninth Street in Manhattan, struggling to make a name for himself. He worked off-Broadway, sometimes far off, and in television commercials, taking all the steps he could to get noticed, build a career, and earn the money to eat an occasional hanger steak in a restaurant without having to duck out before the check arrived. And like many other wannabes, no matter how hard this aspiring actor worked to get the right parts, make the right career choices, and excel in his trade, his most reliable role remained the one he played in his other career—as a bartender. Then one day in the summer of 1984 he flew to Los Angeles, either to attend the Olympics (if you believe his publicist) or to visit a girlfriend (if you believe *The New York Times*). Whichever account is accurate, one thing is clear: the decision to visit the West Coast had little to do with acting and much to do with love, or at least the love of sports. Yet it proved to be the best career decision he ever made, most likely the best decision of his life.

The actor's name is Bruce Willis, and while he was in Los Angeles, an agent suggested he go on a few television auditions.[11] One was for a show in its final stages of casting. The producers already had a list of finalists in mind, but in Hollywood nothing is final until the ink on the contract is dry and the litigation has ended. Willis got his audition and landed the role—that of David Addison, the male lead paired with Cybill Shepherd in a new ABC offering called *Moon-lighting*.

It might be tempting to believe that Willis was the obvious choice over Mr. X, the fellow at the top of the list when the newcomer arrived, and that the rest is, as they say, history. Since in hindsight we know that *Moonlighting* and Willis became huge successes, it is hard to imagine the assemblage of Hollywood decision makers, upon seeing Willis, doing anything but lighting up stogies as they celebrated their brilliant discovery and put to flame their now-outmoded list of finalists. But what really happened at the casting session is more like what you get when you send your kids out for a single gallon of ice cream and two want strawberry while the third demands triple-chocolate-fudge brownie. The network executives fought for Mr. X, their judgment being that Willis did not look like a serious lead. Glenn Caron, *Moonlighting*'s executive producer, argued for Willis. It is easy, looking back, to dismiss the network executives as ignorant buffoons. In my experience, television producers often do, especially when the executives are out of earshot. But before we make that choice, consider this: television viewers at first agreed with the executives' mediocre assessment. *Moonlighting* debuted in March 1985 to low ratings and continued with a mediocre performance through the rest of its first season. Only in the second season did viewers change their minds and the show become a major hit. Willis's appeal and his success were apparently unforeseeable until, of course, he suddenly became a star. One might at this point chalk up the story to crazy Hollywood, but Willis's drunkard's walk to success is not at all unusual. A path punctuated by random impacts and unintended consequences is the path of many successful people, not only in their careers but also in their loves, hobbies, and friendships. In fact, it is more the rule than the exception.

I was watching late-night television recently when another star, though not one from the entertainment world, appeared for an interview. His name is Bill Gates. Though the interviewer is known for his sarcastic approach, toward Gates he seemed unusually deferential. Even the audience seemed to ogle Gates. The reason, of course, is that for thirteen years straight Gates was named the richest man in the world by *Forbes* magazine. In fact, since founding Microsoft, Gates has earned more than $100 a second. And so when he was asked about his vision for interactive television, everyone waited with great anticipation to hear what he had to say. But his answer was ordinary, no more creative, ingenious, or insightful than anything I've heard from a dozen other computer professionals. Which brings us to this question: does Gates earn $100 per second because he is god-like, or is he godlike because he earns $100 per second?

In August 1980, when a group of IBM employees working on a secret project to build a personal computer flew to Seattle to meet with the young computer entrepreneur, Bill Gates was running a small company and IBM needed a program, called an operating system, for its planned "home computer." Recollections of the ensuing events vary, but the gist goes like this:[12] Gates said he couldn't provide the operating system and referred the IBM people to a famed programmer named Gary Kildall at Digital Research Inc. The talks between IBM and Kildall did not go well. For one thing, when IBM showed up at DRI's offices, Kildall's then wife, the company's business manager, refused to sign IBM's nondisclosure agreement. The IBM emissaries called again, and that time Kildall did meet with them. No one knows exactly what transpired in that meeting, but if an informal deal was made, it didn't stick. Around that time one of the IBM employees, Jack Sams, saw Gates again. They both knew of another operating system that was available, a system that was, depending on whom you ask, based on or inspired by Kildall's. According to Sams, Gates said, "Do you want to get . . . [that operating system], or do you want me to?" Sams, apparently not appreciating the implications, said, "By all means, you get it." Gates did, for $50,000 (or, by some accounts, a bit more), made a few changes, and renamed it DOS (disk operating system). IBM, apparently with little faith in the potential of its new idea, licensed DOS from Gates for a low per-copy royalty fee, letting Gates retain the rights. DOS was no better—and many, including most computer professionals, would claim far worse—than, say, Apple's Macintosh operating system. But the growing base of IBM users encouraged software developers to write for DOS, thereby encouraging prospective users to buy IBM machines, a circumstance that in turn encouraged software developers to write for DOS. In other words, as W. Brian Arthur would say, people bought DOS because people were buying DOS. In the fluid world of computer entrepreneurs, Gates became the molecule that broke from the pack. But had it not been for Kildall's uncooperativeness, IBM's lack of vision, or the second encounter between Sams and Gates, despite whatever visionary or business acumen Gates possesses, he might have become just another software entrepreneur rather than the richest man in the world, and that is probably why his vision seems like that of just that—another software entrepreneur.

Our society can be quick to make wealthy people into heroes and poor ones into goats. That's why the real estate mogul Donald Trump, whose Plaza Hotel went bankrupt and whose casino empire went bankrupt twice (a shareholder who invested $10,000 in his casino company in 1994 would thirteen years later have come away with $636),[13] nevertheless dared to star in a wildly successful television program in which he judged the business acumen of aspiring young people.

Obviously it can be a mistake to assign brilliance in proportion to wealth. We cannot see a person's potential, only his or her results, so we often misjudge people by thinking that the results must reflect the person. The normal accident theory of life shows not that the connection between actions and rewards is random but that random influences are as important as our qualities and actions.

On an emotional level many people resist the idea that random influences are important even if, on an intellectual level, they understand that they are. If people underestimate the role of chance in the careers of moguls, do they also downplay its role in the lives of the least successful? In the 1960s that question inspired the social psychologist Melvin Lerner to look into society's negative attitudes toward the poor.[14] Realizing that "few people would engage in extended activity if they believed that there were a random connection between what they did and the rewards they received,"[15] Lerner concluded that "for the sake of their own sanity," people overestimate the degree to which ability can be inferred from success.[16] We are inclined, that is, to see movie stars as more talented than aspiring movie stars and to think that the richest people in the world must also be the smartest.

WE MIGHT NOT THINK we judge people according to their income or outward signs of success, but even when we know for a fact that a person's salary is completely random, many people cannot avoid making the intuitive judgment that salary is correlated with worth. Melvin Lerner examined that issue by arranging for subjects to sit in a small darkened auditorium facing a one-way mirror.[17] From their seats these observers could gaze into a small well-lit room containing a table and two chairs. The observers were led to believe that two workers, Tom and Bill, would soon enter the room and work together for fifteen minutes unscrambling anagrams. The curtains in front of the viewing window were then closed, and Lerner told the observers that he would keep the curtains shut because the experiment would go better if they could hear but not see the workers, so that they would not be influenced by their appearance. He also told them that because his funds were limited, he could pay only one of the workers, who would be chosen at random. As Lerner left the room, an assistant threw a switch that began to play an audiotape. The observers believed they were listening in as Tom and Bill entered the room behind the curtain and began their work. Actually they were listening to a recording of Tom and Bill reading a fixed script, which had been constructed such that, by various objective measures, each of them appeared to be equally adept and successful at his task. Afterward the observers, not knowing this, were asked to rate Tom and Bill on their effort, creativity, and success. When Tom was selected to receive the payment, about 90 percent of the observers rated him as having made the greater contribution. When Bill was selected, about 70 percent of the observers rated him higher. Despite Tom and Bill's equivalent performance and the observers' knowledge that the pay was randomly assigned, the observers perceived the worker who got paid as being better than the one who had worked for nothing. Alas, as all of those who dress for success know, we are all too easily fooled by the money someone earns.

A series of related studies investigated the same effect from the point of view of the workers themselves.[18] Everyone knows that bosses with the right social and academic credentials and a nice title and salary have at times put a higher value on their own ideas than on those of their underlings. Researchers wondered, will people who earn more money purely by chance behave the same way? Does even unearned "success" instill a feeling of superiority? To find out, pairs of volunteers were asked to cooperate on various pointless tasks. In one task, for instance, a black-and-white image was briefly displayed and the subjects had to decide whether the top or the bottom of the image contained a greater proportion of white. Before each task began, one of the subjects was randomly chosen to receive considerably more pay for participating than the other. When that information was not made available, the subjects cooperated pretty harmoniously. But when they knew how much they each were getting paid, the higher-paid subjects exhibited more resistance to input from their partners than the lower-paid ones. Even random differences in pay lead to the backward inference of differences in skill and hence to the development of unequal influence. It's an element of personal and office dynamics that cannot be ignored.

But it is the other side of the question that was closer to the original motivation for Lerner's work. With a colleague, Lerner asked whether people are inclined to feel that those who are not successful or who suffer deserve their fate.[19] In that study small groups of female college students gathered in a waiting room. After a few minutes one was selected and led out. That student, whom I will call the victim, was not really a test subject but had been planted in the room by the experimenters. The remaining subjects were told that they would observe the victim working at a learning task and that each time she made an incorrect response, she would receive an electric shock. The experimenter adjusted some knobs said to control the shock levels, and then a video monitor was turned on. The subjects watched as the victim entered an adjoining room, was strapped to a "shock apparatus," and then attempted to learn pairs of nonsense syllables.

During the task the victim received several apparently painful electric shocks for her incorrect responses. She reacted with exclamations of pain and suffering. In reality the victim was acting, and what played on the monitor was a prerecorded tape. At first, as expected, most of the observers reported being extremely upset by their peer's unjust suffering. But as the experiment continued, their sympathy for the victim began to erode. Eventually the observers, powerless to help, instead began to denigrate the victim. The more the victim suffered, the lower their opinion of her became. As Lerner had predicted, the observers had a need to understand the situation in terms of cause and effect. To be certain that some other dynamic wasn't really at work, the experiment was repeated with other groups of subjects, who were told that the victim would be well compensated for her pain. In other words, these subjects believed that the victim was being "fairly" treated but were otherwise exposed to the same scenario. Those observers did not develop a tendency to think poorly of the victim. We unfortunately seem to be unconsciously biased against those in society who come out on the bottom.

We miss the effects of randomness in life because when we assess the world, we tend to see what we expect to see. We in effect define degree of talent by degree of success and then reinforce our feelings of causality by noting the correlation. That's why although there is sometimes little difference in ability between a wildly successful person and one who is not as successful, there is usually a big difference in how they are viewed. Before *Moonlighting*, if you were told by the young bartender Bruce Willis that he hoped to become a film star, you would not have thought, *gee, I sure am lucky to have this chance to chat one-on-one with a charismatic future celebrity*, but rather you would have thought something more along the lines of *yeah, well, for now just make sure not to overdo it on the vermouth*. The day after the show became a hit, however, everyone suddenly viewed Bruce Willis as a star, a guy who has that something special it takes to capture viewers' hearts and imagination.

The power of expectations was dramatically illustrated in a bold experiment conducted years ago by the psychologist David L. Rosenhan.[20] In that study each of eight "pseudopatients" made an appointment at one of a variety of hospitals and then showed up at the admissions office complaining that they were hearing strange voices. The pseudopatients were a varied group: three psychologists, a psychiatrist, a pediatrician, a student, a painter, and a housewife. Other than alleging that single symptom and reporting false names and vocations, they all described their lives with complete honesty. Confident in the clockwork operation of our mental health system, some of the subjects later reported having feared that their obvious sanity would be immediately sniffed out, causing great embarrassment on their part. They needn't have worried. All but one were admitted to the hospital with a diagnosis of schizophrenia. The remaining patient was admitted with a diagnosis of manic-depressive psychosis.

Upon admission, they all ceased simulating any symptoms of abnormality and reported that the voices were gone. Then, as previously instructed by Rosenhan, they waited for the staff to notice that they were not, in fact, insane. None of the staff noticed. Instead, the hospital workers interpreted the pseudopatients' behavior through the lens of insanity. When one patient was observed writing in his diary, it was noted in the nursing record that "patient engages in writing behavior," identifying the writing as a sign of mental illness. When another patient had an outburst while being mistreated by an attendant, the behavior was also assumed to be part of the patient's pathology. Even the act of arriving at the cafeteria before it opened for lunch was seen as a symptom of insanity. Other patients, unimpressed by formal medical diagnoses, would regularly challenge the pseudopatients with comments like "You're not crazy. You're a journalist . . . you're checking up on the hospital." The pseudopatients' doctors, however, wrote notes like "This white 39-year-old male . . . manifests a long history of considerable ambivalence in close relationships, which begins in early childhood. A warm relationship with his mother cools during adolescence. A distant relationship with his father is described as being very intense."

The good news is that despite their suspicious habits of writing and showing up early for lunch, the pseudopatients were judged not to be a danger to themselves or others and released after an average stay of nineteen days. The hospitals never detected the ruse and, when later informed of what had gone on, denied that such a scenario could be possible.

If it is easy to fall victim to expectations, it is also easy to exploit them. That is why struggling people in Hollywood work hard to look as though they are not struggling, why doctors wear white coats and place all manner of certificates and degrees on their office walls, why used-car salesmen would rather repair blemishes on the outside of a car than sink money into engine work, and why teachers will, on average, give a higher grade to a homework assignment turned in by an "excellent" student than to identical work turned in by a "weak" one.[21] Marketers also know this and design ad campaigns to create and then exploit our expectations. One arena in which that was done very effectively is the vodka market. Vodka is a neutral spirit, distilled, according to the U.S. government definition, "as to be without distinctive character, aroma, taste or color." Most American vodkas originate, therefore, not with passionate, flannel-shirted men like those who create wines, but with corporate giants like the agrochemical supplier Archer Daniels Midland. And the job of the vodka distiller is not to nurture an aging process that imparts finely nuanced flavor but to take the 190-proof industrial swill such suppliers provide, add water, and *subtract* as much of the taste as possible. Through massive image-building campaigns, however, vodka producers have managed to create very strong expectations of difference. As a result, people believe that this liquor, which by its very *definition* is without a distinctive character, actually varies greatly from brand to brand. Moreover, they are willing to pay large amounts of money based on those differences. Lest I be dismissed as a tasteless boor, I wish to point out that there is a way to test my ravings. You could line up a series of vodkas and a series of vodka sophisticates and perform a blind tasting. As it happens, *The New York Times* did just that.[22] And without their labels, fancy vodkas like Grey Goose and Ketel One didn't fare so well. Compared with conventional wisdom, in fact, the results appeared random. Moreover, of the twenty-one vodkas tasted, it was the cheap bar brand, Smirnoff, that came out at the top of the list. Our assessment of the world would be quite different if all our judgments could be insulated from expectation and based only on relevant data.

A FEW YEARS AGO *The Sunday Times* of London conducted an experiment. Its editors submitted typewritten manuscripts of the opening chapters of two novels that had won the Booker Prize—one of the world's most prestigious and most influential awards for contemporary fiction—to twenty major publishers and agents.[23] One of the novels was *In a Free State* by V. S. Naipaul, who won the Nobel Prize for Literature; the other was *Holiday* by Stanley Middleton. One can safely assume that each of the recipients of the manuscripts would have heaped praise on the highly lauded novels had they known what they were reading. But the submissions were made as if they were the work of aspiring authors, and none of the publishers or agents appeared to recognize them. How did the highly successful works fare? All but one of the replies were rejections. The exception was an expression of interest in Middleton's novel by a London literary agent. The same agent wrote of Naipaul's book, "We . . . thought it was quite original. In the end though I'm afraid we just weren't quite enthusiastic enough to be able to offer to take things further."

The author Stephen King unwittingly conducted a similar experiment when, worried that the public would not accept his books as quickly as he could churn them out, he wrote a series of novels under the pseudonym Richard Bachman. Sales figures indicated that even Stephen King, without the name, is no Stephen King. (Sales picked up considerably after word of the author's true identity finally got out.) Sadly, one experiment King did not perform was the opposite:

to swathe wonderful unpublished manuscripts by struggling writers in covers naming him as the author. But if even Stephen King, without the name, is no Stephen King, then the rest of us, when our creative work receives a less-than-Kingly reception, might take comfort in knowing that the differences in quality might not be as great as some people would have us believe.

Years ago at Caltech, I had an office around the corner from the office of a physicist named John Schwarz. He was getting little recognition and had suffered a decade of ridicule as he almost single-handedly kept alive a discredited theory, called string theory, which predicted that space has many more dimensions than the three we observe. Then one day he and a co-worker made a technical breakthrough, and for reasons that need not concern us here, suddenly the extra dimensions sounded more acceptable. String theory has been the hottest thing in physics ever since. Today John is considered one of the brilliant elder statesmen of physics, yet had he let the years of obscurity get to him, he would have been a testament to Thomas Edison's observation that "many of life's failures are people who did not realize how close they were to success when they gave up."[24]

Another physicist I knew had a story that was strikingly similar to John's. He was, coincidentally, John's PhD adviser at the University of California, Berkeley. Considered one of the most brilliant scientists of his generation, this physicist was a leader in an area of research called S-matrix theory. Like John, he was stubbornly persistent and continued to work on his theory for years after others had given up. But unlike John, he did not succeed. And because of his lack of success he ended his career with many people thinking him a crackpot. But in my opinion both he and John were brilliant physicists with the courage to work—with no promise of an imminent breakthrough— on a theory that had gone out of style. And just as authors should be judged by their writing and not their books' sales, so physicists—and all who strive to achieve—should be judged more by their abilities than by their success.

The cord that tethers ability to success is both loose and elastic. It is easy to see fine qualities in successful books or to see unpublished manuscripts, inexpensive vodkas, or people struggling in any field as somehow lacking. It is easy to believe that ideas that worked were good ideas, that plans that succeeded were well designed, and that ideas and plans that did not were ill conceived. And it is easy to make heroes out of the most successful and to glance with disdain at the least. But ability does not guarantee achievement, nor is achievement proportional to ability. And so it is important to always keep in mind the other term in the equation—the role of chance.

It is no tragedy to think of the most successful people in any field as superheroes. But it is a tragedy when a belief in the judgment of experts or the marketplace rather than a belief in ourselves causes us to give up, as John Kennedy Toole did when he committed suicide after publishers repeatedly rejected his manuscript for the posthumously best-selling *Confederacy of Dunces*. And so when tempted to judge someone by his or her degree of success, I like to remind myself that were they to start over, Stephen King might be only a Richard Bachman and V. S. Naipaul just another struggling author, and somewhere out there roam the equals of Bill Gates and Bruce Willis and Roger Maris who are not rich and famous, equals on whom Fortune did not bestow the right breakthrough product or TV show or year. What I've learned, above all, is to keep marching forward because the best news is that since chance does play a role, one important factor in success *is* under our control: the number of at bats, the number of chances taken, the number of opportunities seized. For even a coin weighted toward failure will sometimes land on success. Or as the IBM pioneer Thomas Watson said, "If you want to succeed, double your failure rate."

I have tried in this book to present the basic concepts of randomness, to illustrate how they apply to human affairs, and to present my view that its effects are largely overlooked in our interpretations of events and in our expectations and decisions. It may come as an epiphany merely to recognize the ubiquitous role of random processes in our lives; the true power of the theory of random processes, however, lies in the fact that once we understand the nature of random processes, we can alter the way we perceive the events that happen around us.

The psychologist David Rosenhan wrote that "once a person is abnormal, all of his other behaviors and characteristics are colored by that label."[25] The same applies for stardom, for many other labels of success, and for those of failure. We judge people and initiatives by their results, and we expect events to happen for good, understandable reasons. But our clear visions of inevitability are often only illusions. I wrote this book in the belief that we can reorganize our thinking in the face of uncertainty. We can improve our skill at decision making and tame some of the biases that lead us to make poor judgments and poor choices. We can seek to understand people's qualities or the qualities of a situation quite apart from the results they attain, and we can learn to judge decisions by the spectrum of potential outcomes they might have produced rather than by the particular result that actually occurred.

My mother always warned me not to think I could predict or control the future. She once related the incident that converted her to that belief. It concerned her sister, Sabina, of whom she still often speaks although it has been over sixty-five years since she last saw her. Sabina was seventeen. My mother, who idolized her as younger siblings sometimes do their older siblings, was fifteen. The Nazis had invaded Poland, and my father, from the poor section of town, had joined the underground and, as I said earlier, eventually ended up in Buchenwald. My mother, who didn't know him then, came from the wealthy part of town and ended up in a forced-labor camp. There she was given the job of nurse's aide and took care of patients suffering from typhus. Food was scarce, and random death was always near. To help protect my mother from the ever-present dangers, Sabina agreed to a plan. She had a friend who was a member of the Jewish police, a group, generally despised by the inmates, who carried out the Germans' commands and helped keep order in the camp. Sabina's friend had offered to marry her—a marriage in name only— so that Sabina might obtain the protections that his position afforded.

Sabina, thinking those protections would extend to my mother, agreed. For a while it worked. Then something happened, and the Nazis soured on the Jewish police. They sent a number of officers to the gas chambers, along with their spouses—including Sabina's husband and Sabina herself. My mother has lived now for many more years without Sabina than she had lived with her, but Sabina's death still haunts her. My mother worries that when she is gone, there will no longer be any trace that Sabina ever existed. To her this story shows that it is pointless to make plans. I do not agree. I believe it is important to plan, if we do so with our eyes open. But more important, my mother's experience has taught me that we ought to identify and appreciate the good luck that we have and recognize the random events that contribute to our success. It has taught me, too, to accept the chance events that may cause us grief. Most of all it has taught me to appreciate the absence of bad luck, the absence of events that might have brought us down, and the absence of the disease, war, famine, and accident that have not—or have not yet—befallen us.

ACKNOWLEDGMENTS

I ASSUME if you are reading this far, that you liked this book. For its good qualities, I'd like to claim all credit, but as Nixon once said, that would be wrong. And so I'd like to point out the people who, with their time, knowledge, talent, and/or patience, helped me to create a book that is better than any which I could have created alone. First, Donna Scott, Mark Hillery, and Matt Costello gave me constant encouragement. Mark, in particular, wanted me to write a book about entropy, but then listened (and read) patiently as I instead applied many of those same ideas to the everyday world. My agent, Susan Ginsburg, never wanted me to write a book about entropy, but, like Mark, was a source of unwaivering constructive input and encouragement. My friend Judith Croasdell was always supportive, and, when called upon, also worked a miracle or two. And my editor, Edward Kastenmeier, never grew tired of the long discussions I drew him into about the style and content of virtually every sentence, or more likely, was too polite to complain about it. I also owe a debt to Edward's colleagues, Marty Asher, Dan Frank, and Tim O'Connell, who, along with Edward, nurtured this work and helped shape the text and to Janice Goldklang, Michiko Clark, Chris Gillespie, Keith Goldsmith, James Kimball, and Vannessa Schneider whose tireless efforts helped get this to you.

On the technical side, Larry Goldstein and Ted Hill inspired me in numerous fun and exciting mathematical debates and discussions, and gave me invaluable feedback on the manuscript. Fred Rose seemed to have left his job at *The Wall Street Journal* solely to free up time to lend me advice on the workings of the financial markets. Lyle Long applied his considerable expertise at data analysis to help create the graphs related to fund manager performance. And Christof Koch welcomed me into his lab at Caltech and opened my eyes to the exciting new developments in neuroscience that pepper these pages. Many other friends and colleagues read chapters, sometimes more than one draft, or otherwise provided useful suggestions or information. They include Jed Buchwald, Lynne Cox, Richard Cheverton, Rebecca Forster, Miriam Goodman, Catherine Keefe, Jeff Mackowiak, Cindy Mayer, Patricia McFall, Andy Meisler, Steve Mlodinow, Phil Reed, Seth Roberts, Laura Saari, Matt Salganik, Martin Smith, Steve Thomas, Diane Turner, and Jerry Webman. Thanks to you all. Finally, I owe a profound thank you to my family, Donna, Alexei, Nicolai, Olivia, and to my mother, Irene, from each of whom I often stole time in order that I might improve upon, or at least obsess over, this work.

NOTES

Prologue

1. Stanley Meisler, "First in 1763: Spain Lottery—Not Even War Stops It," *Los Angeles Times*, December 30, 1977.

2. On basketball, see Michael Patrick Allen, Sharon K. Panian, and Roy E. Lotz, "Managerial Succession and Organizational Performance: A Recalcitrant Problem Revisited," *Administrative Science Quarterly* 24, no. 2 (June 1979): 167–80; on football, M. Craig Brown, "Administrative Succession and Organizational Performance: The Succession Effect," *Administrative Science Quarterly* 27, no. 1 (March 1982): 1–16; on baseball, Oscar Grusky, "Managerial Succession and Organizational Effectiveness," *American Journal of Sociology* 69, no. 1 (July 1963): 21–31, and William A. Gamson and Norman A. Scotch, "Scapegoating in Baseball," *American Journal of Sociology* 70, no. 1 (July 1964): 69–72; on soccer, Ruud H. Koning, "An Econometric Evaluation of the Effect of Firing a Coach on Team Performance," *Applied Economics* 35, no. 5 (March 2003): 555–64.

3. James Surowiecki, *The Wisdom of Crowds* (New York: Doubleday, 2004), pp. 218–19.

4. Armen Alchian, "Uncertainty, Evolution, and Economic Theory," *Journal of Political Economy* 58, no. 3 (June 1950): 213.

Chapter 1: Peering through the Eyepiece of Randomness

1. Kerstin Preuschoff, Peter Bossaerts, and Steven R. Quartz, "Neural Differentiation of Expected Reward and Risk in Human Subcortical Structures," *Neuron* 51 (August 3, 2006): 381–90.

2. Benedetto De Martino et al., "Frames, Biases, and Rational Decision-Making in the Human Brain," *Science* 313 (August 4, 2006): 684–87.

3. George Wolford, Michael B. Miller, and Michael Gazzaniga, "The Left Hemisphere's Role in Hypothesis Formation," *Journal of Neuroscience* 20:RC64 (2000): 1–4.

4. Bertrand Russell, *An Inquiry into Meaning and Truth* (1950; repr., Oxford: Routledge, 1996), p. 15.

5. Matt Johnson and Tom Hundt, "Hog Industry Targets State for Good Reason," *Vernon County (Wisconsin) Broadcaster*, July 17, 2007.

6. Kevin McKean, "Decisions, Decisions," *Discover,* June 1985, pp. 22–31.

7. David Oshinsky, "No Thanks, Mr. Nabokov," *New York Times Book Review,* September 9, 2007.

8. Press accounts of the number of rejections these manuscripts received vary slightly.

9. William Goldman, *Adventures in the Screen Trade* (New York: Warner Books, 1983), p. 41.

10. See Arthur De Vany, *Hollywood Economics* (Abington, U.K.: Routledge, 2004).

11. William Feller, *An Introduction to Probability Theory and Its Applications,* 2nd ed. (New York: John Wiley and Sons, 1957), p. 68. Note that for simplicity's sake, when the opponents are tied, Feller defines the lead as belonging to the player who led at the preceding trial.

12. Leonard Mlodinow, "Meet Hollywood's Latest Genius," *Los Angeles Times Magazine,* July 2, 2006.

13. Dave McNary, "Par Goes for Grey Matter," *Variety,* January 2, 2005.

14. Ronald Grover, "Paramount's Cold Snap: The Heat Is On," *BusinessWeek,* November 21, 2003.

15. Dave McNary, "Parting Gifts: Old Regime's Pics Fuel Paramount Rebound," *Variety,* August 16, 2005.

16. Anita M. Busch, "Canton Inks Prod'n Pact at Warner's," *Variety,* August 7, 1997.

17. "The Making of a Hero," *Time,* September 29, 1961, p. 72. The old-timer was Rogers Hornsby.

18. "Mickey Mantle and Roger Maris: The Photographic Essay," *Life,* August 18, 1961, p. 62.

19. For those who don't know baseball, the plate is a rubber slab embedded in the ground, which a player stands before as he attempts to hit the ball. For those who do know baseball, please note that I included walks in my definition of opportunities. If the calculation is redone employing just official at bats, the result is about the same.

20. See Stephen Jay Gould, "The Streak of Streaks," *New York Review of Books,* August 18, 1988, pp. 8–12 (we'll come back to their work in more detail later). A compelling and mathematically detailed analysis of coin-toss models in sports appears in chapter 2 of a book in progress by Charles M. Grinstead, William P. Peterson, and J. Laurie Snell, tentatively titled *Fat Chance;* www.math.dartmouth.edu/~prob/prob/NEW/bestofchance.pdf.

Chapter 2: The Laws of Truths and Half-Truths

1. Daniel Kahneman, Paul Slovic, and Amos Tversky, eds., *Judgment under Uncertainty: Heuristics and Biases* (Cambridge: Cambridge University Press, 1982), pp. 90–98.

2. Amos Tversky and Daniel Kahneman, "Extensional versus Intuitive Reasoning: The Conjunction Fallacy in Probability Judgment," *Psychological Review* 90, no. 4 (October 1983): 293–315.

3. Craig R. Fox and Richard Birke, "Forecasting Trial Outcomes: Lawyers Assign Higher Probabilities to Possibilities That Are Described in Greater Detail," *Law and Human Behavior* 26, no. 2 (April 2002): 159–73.

4. Plato, *The Dialogues of Plato,* trans. Benjamin Jowett (Boston: Colonial Press, 1899), p. 116.

5. Plato, *Theaetetus* (Whitefish, Mont.: Kessinger, 2004), p. 25.

6. Amos Tversky and Daniel Kahneman, "Availability: A Heuristic for Judging Frequency and Probability," *Cognitive Psychology* 5 (1973): 207–32.

7. Reid Hastie and Robyn M. Dawes, *Rational Choice in an Uncertain World: The Psychology and Judgement of Decision Making* (Thousand Oaks, Calif.: Sage, 2001), p. 87.

8. Robert M. Reyes, William C. Thompson, and Gordon H. Bower, "Judgmental Biases Resulting from Differing Availabilities of Arguments," *Journal of Personality and Social Psychology* 39, no. 1 (1980): 2–12.

9. Robert Kaplan, *The Nothing That Is: A Natural History of Zero* (London: Oxford University Press, 1999), pp. 15–17.

10. Cicero, quoted in Morris Kline, *Mathematical Thought from Ancient to Modern Times* (London: Oxford University Press, 1972), 1:179.

11. Morris Kline, *Mathematics in Western Culture* (London: Oxford University Press, 1953), p. 86.

12. Kline, *Mathematical Thought,* pp. 178–79.

13. Cicero, quoted in Warren Weaver, *Lady Luck* (Mineola, N.Y.: Dover Publications, 1982), p. 53.

14. Cicero, quoted in F. N. David, *Gods, Games and Gambling: A History of Probability and Statistical Ideas* (Mineola, N.Y.: Dover Publications, 1998), pp. 24–26.

15. Cicero, quoted in Bart K. Holland, *What Are the Chances? Voodoo Deaths, Office Gossip, and Other Adventures in Probability* (Baltimore: Johns Hopkins University Press, 2002), p. 24.

16. Ibid., p. 25.

17. James Franklin, *The Science of Conjecture: Evidence and Probability before Pascal* (Baltimore: Johns Hopkins University Press, 2001), pp. 4, 8.

18. Quoted ibid., p. 13.

19. Quoted ibid., p. 14.

20. William C. Thompson, Franco Taroni, and Colin G. G. Aitken, "How the Probability of a False Positive Affects the Value of DNA Evidence," *Journal of Forensic Sciences* 48, no. 1 (January 2003): 1–8.

21. Ibid., p. 2. The story is recounted in Bill Braun, "Lawyers Seek to Overturn Rape Conviction," *Tulsa World,* November 22, 1996. See also www.innocenceproject .org. (Durham was released in 1997.)

22. *People v. Collins*, 68 Calif. 2d 319, 438 P.2d 33, 66 Cal. Rptr. 497 (1968).

23. Thomas Lyon, private communication.

Chapter 3: Finding Your Way through a Space of Possibilities

1. Alan Wykes, *Doctor Cardano: Physician Extraordinary* (London: Frederick Muller, 1969). See also Oystein Ore, *Cardano: The Gambling Scholar*, with a translation of Cardano's *Book on Games of Chance* by Sydney Henry Gould (Princeton, N.J.: Princeton University Press, 1953).

2. Marilyn vos Savant, "Ask Marilyn," *Parade*, September 9, 1990.

3. Bruce D. Burns and Mareike Wieth, "Causality and Reasoning: The Monty Hall Dilemma," in *Proceedings of the Twenty-fifth Annual Meeting of the Cognitive Science Society*, ed. R. Alterman and D. Kirsh (Hillsdale, N.J.: Lawrence Erlbaum Associates, 2003), p. 198.

4. National Science Board, *Science and Engineering Indicators—2002* (Arlington, Va.: National Science Foundation, 2002); http://www.nsf.gov/statistics/seind02/. See vol. 2, chap. 7, table 7-10.

5. Gary P. Posner, "Nation's Mathematicians Guilty of Innumeracy," *Skeptical Inquirer* 15, no. 4 (Summer 1991).

6. Bruce Schechter, *My Brain Is Open: The Mathematical Journeys of Paul Erdös* (New York: Touchstone, 1998), pp. 107–9.

7. Ibid., pp. 189–90, 196–97.

8. John Tierney, "Behind Monty's Doors: Puzzle, Debate and Answer?" *New York Times*, July 21, 1991.

9. Robert S. Gottfried, *The Black Death: Natural and Human Disaster in Medieval Europe* (New York: Free Press, 1985).

10. Gerolamo Cardano, quoted in Wykes, *Doctor Cardano*, p. 18.

11. Kline, *Mathematical Thought*, pp. 184–85, 259–60.

12. "Oprah's New Shape: How She Got It," *O, the Oprah Magazine*, January 2003.

13. Lorraine J. Daston, *Classical Probability in the Enlightenment* (Princeton, N.J.: Princeton University Press, 1998), p. 97.

14. Marilyn vos Savant, "Ask Marilyn," *Parade*, March 3, 1996, p. 14.

15. There are four tires on the car, so, letting RF signify "right front," and so on, there are 16 possible combinations of responses by the two students. If the first response listed represents that of student 1 and the second that of student 2, the possible joint responses are (RF, RF), (RF, LF), (RF, RR), (RF, LR), (LF, RF), (LF, LF), (LF, RR), (LF, LR), (RR, RF), (RR, LF), (RR, RR), (RR, LR), (LR, RF), (LR, LF), (LR, RR), (LR, LR). Of these 16, 4 are in agreement: (RF, RF), (LF, LF), (RR, RR), (LR, LR). Hence the chances are 4 out of 16, or 1 in 4.

16. Martin Gardner, "Mathematical Games," *Scientific American*, October 1959, pp. 180–82.

17. Jerome Cardan, *The Book of My Life: De Vita Propia Liber*, trans. Jean Stoner (Whitefish, Mont.: Kessinger, 2004), p. 35.

18. Cardano, quoted in Wykes, *Doctor Cardano*, p. 57.

19. Cardano, quoted ibid.

20. Cardano, quoted ibid., p. 172.

Chapter 4: Tracking the Pathways to Success

1. Bengt Ankarloo and Stuart Clark, eds., *Witchcraft and Magic in Europe: The Period of the Witch Trials* (Philadelphia: University of Pennsylvania Press, 2002), pp. 99–104.

2. Meghan Collins, "Traders Ward Off Evil Spirits," October 31, 2003, http://www.CNNMoney.com/2003/10/28/markets_trader_superstition/index.htm.

3. Henk Tijms, *Understanding Probability: Chance Rules in Everyday Life* (Cambridge: Cambridge University Press, 2004), p. 16.

4. Ibid., p. 80.

5. David, *Gods, Games and Gambling*, p. 65.

6. Blaise Pascal, quoted in Jean Steinmann, *Pascal*, trans. Martin Turnell (New York: Harcourt, Brace & World, 1962), p. 72.

7. Gilberte Pascal, quoted in Morris Bishop, *Pascal: The Life of a Genius* (1936; repr., New York: Greenwood Press, 1968), p. 47.

8. Ibid., p. 137.

9. Gilberte Pascal, quoted ibid., p. 135.

10. See A.W.F. Edwards, *Pascal's Arithmetical Triangle: The Story of a Mathematical Idea* (Baltimore: Johns Hopkins University Press, 2002).

11. Blaise Pascal, quoted in Herbert Westren Turnbull, *The Great Mathematicians* (New York: New York University Press, 1961), p. 131.

12. Blaise Pascal, quoted in Bishop, *Pascal*, p. 196.

13. Blaise Pascal, quoted in David, *Gods, Games and Gambling*, p. 252.

14. Bruce Martin, "Coincidences: Remarkable or Random?" *Skeptical Inquirer* 22, no. 5 (September/October 1998).

15. Holland, *What Are the Chances?* pp. 86–89.

Chapter 5: The Dueling Laws of Large and Small Numbers

1. Tijms, *Understanding Probability*, p. 53.

2. Scott Kinney, "Judge Sentences Kevin L. Lawrence to 20 Years Prison in Znetix/HMC Stock Scam," Washington State Department of Financial Institutions, press release, November 25, 2003; http://www.dfi.wa.gov/sd/kevin_laurence_sentence.htm.

3. Interview with Darrell Dorrell, August 1, 2005.

4. Lee Berton, "He's Got Their Number: Scholar Uses Math to Foil Financial Fraud," *Wall Street Journal*, July 10, 1995.

5. Charles Sanders Peirce, Max Harold Fisch, and Christian J. W. Kloesel, *Writings of Charles S. Peirce: A Chronological Edition* (Bloomington: Indiana University Press, 1982), p. 427.

6. Rand Corporation, *A Million Random Digits with 100,000 Normal Deviates* (1955; repr., Santa Monica, Calif.: Rand, 2001), pp. ix–x. See also Lola L. Lopes, "Doing the Impossible: A Note on Induction and the Experience of Randomness," *Journal of Experimental Psychology: Learning, Memory, and Cognition* 8, no. 6 (November 1982): 626–36.

7. The account of Joseph Jagger (sometimes spelled Jaggers) is from John Grochowski, "House Has a Built-in Edge When Roulette Wheel Spins," *Chicago Sun-Times*, February 21, 1997.

8. For details about the Bernoulli family and Jakob's life, see E. S. Pearson, ed., *The History of Statistics in the 17th and 18th Centuries against the Changing Background of Intellectual, Scientific and Religious Thought: Lectures by Karl Pearson Given at University College, London, during the Academic Sessions 1921–1933* (New York: Macmillan, 1978), pp. 221–37; J. O. Fleckenstein, "Johann und Jakob Bernoulli," in *Elemente der Mathematik, Beihefte zur Zeitschrift*, no. 6 (Basel, 1949); and Stephen Stigler, "The Bernoullis of Basel," *Journal of Econometrics* 75, no. 1 (1996): 7–13.

9. Quoted in Pearson, *The History of Statistics in the 17th and 18th Centuries*, p. 224.

10. Stephen Stigler, *The History of Statistics: The Measurement of Uncertainty before 1900* (Cambridge, Mass.: Harvard University Press, 1986), p. 65.

11. Pearson, *The History of Statistics in the 17th and 18th Centuries*, p. 226.

12. William H. Cropper, *The Great Physicists: The Life and Times of Leading Physicists from Galileo to Hawking* (London: Oxford University Press, 2001), p. 31.

13. Johann Bernoulli, quoted in Pearson, *The History of Statistics in the 17th and 18th Centuries*, p. 232.

14. This depends, of course, on what you identify as "the modern concept." I am using the definition employed by Hankel's 1871 history of the topic, described in great detail in Gert Schubring, *Conflicts between Generalization, Rigor, and Intuition: Number Concepts Underlying the Development of Analysis in 17th–19th Century France and Germany* (New York: Springer, 2005), pp. 22–32.

15. David Freedman, Robert Pisani, and Roger Purves, *Statistics*, 3rd ed. (New York: W. W. Norton, 1998), pp. 274–75.

16. The Hacking quote is from Ian Hacking, *The Emergence of Probability* (Cambridge: Cambridge University Press, 1975), p. 143. The Bernoulli quote is from David, *Gods, Games and Gambling*, p. 136.

17. For a discussion of what Bernoulli actually proved, see Stigler, *The History of Statistics*, pp. 63–78, and Ian Hacking, *The Emergence of Probability*, pp. 155–65.

18. Amos Tversky and Daniel Kahneman, "Belief in the Law of Small Numbers," *Psychological Bulletin* 76, no. 2 (1971): 105–10.

19. Jakob Bernoulli, quoted in L. E. Maistrov, *Probability Theory: A Historical Sketch*, trans. Samuel Kotz (New York: Academic Press, 1974), p. 68.

20. Stigler, *The History of Statistics*, p. 77.

21. E. T. Bell, *Men of Mathematics* (New York: Simon & Schuster, 1937), p. 134.

Chapter 6: False Positives and Positive Fallacies

1. The account of the Harvard student is from Hastie and Dawes, *Rational Choice in an Uncertain World*, pp. 320–21.

2. I was told a variant of this problem by Mark Hillery of the Physics Department at Hunter College, City University of New York, who heard it while on a trip to Bratislava, Slovakia.

3. Quoted in Stigler, *The History of Statistics*, p. 123.

4. Ibid., pp. 121–31.

5. U.S. Social Security Administration, "Popular Baby Names: Popular Names by Birth Year; Popularity in 1935," http://www.ssa.gov/cgi-bin/popularnames.cgi.

6. Division of HIV/AIDS, Center for Infectious Diseases, *HIV/AIDS Surveillance Report* (Atlanta: Centers for Disease Control, January 1990). I calculated the statistic quoted from the data given but also had to use some estimates. In particular, the data quoted refers to AIDS cases, not HIV infection, but that suffices for the purpose of illustrating the concept.

7. To be precise, the probability that A will occur *if* B occurs is equal to the probability that B will occur if A occurs multiplied by a correction factor that equals the ratio of the probability of A to the probability of B.

8. Gerd Gigerenzer, *Calculated Risks: How to Know When Numbers Deceive You* (New York: Simon & Schuster, 2002), pp. 40–44.

9. Donald A. Barry and LeeAnn Chastain, "Inferences About Testosterone Abuse Among Athletes," *Chance* 17, no. 2 (2004): 5–8.

10. John Batt, *Stolen Innocence* (London: Ebury Press, 2005).

11. Stephen J. Watkins, "Conviction by Mathematical Error? Doctors and Lawyers Should Get Probability Theory Right," *BMJ* 320 (January 1, 2000): 2–3.

12. "Royal Statistical Society Concerned by Issues Raised in Sally Clark Case," Royal Statistical Society, London, news release, October 23, 2001; http://www.rss.org.uk/PDF/RSS%20Statement%20regarding%20statistical%20issues%20in%20the%20Sally%20Clark%20case,%20October%2023rd%202001.pdf.

13. Ray Hill, "Multiple Sudden Infant Deaths—Coincidence or beyond Coincidence?" *Paediatric and Perinatal Epidemiology* 18, no. 5 (September 2004): 320–26.

14. Quoted in Alan Dershowitz, *Reasonable Doubts: The Criminal Justice System and the O. J. Simpson Case* (New York: Simon & Schuster, 1996), p. 101.

15. Federal Bureau of Investigation, "Uniform Crime Reports," http://www.fbi.gov/ucr/ucr.htm.

16. Alan Dershowitz, *The Best Defense* (New York: Vintage, 1983), p. xix.

17. Pierre-Simon de Laplace, quoted in James Newman, ed., *The World of Mathematics* (Mineola, N.Y.: Dover Publications, 1956): 2:1323.

Chapter 7: Measurement and the Law of Errors

1. Sarah Kershaw and Eli Sanders, "Recounts and Partisan Bickering Bring Election Fatigue to Washington Voters," *New York Times*, December 26, 2004; and Timothy Egan, "Trial for Governor's Seat Set to Start in Washington," *New York Times*, May 23, 2005.

2. Jed Z. Buchwald, "Discrepant Measurements and Experimental Knowledge in the Early Modern Era," *Archive for History of Exact Sciences* 60, no. 6 (November 2006): 565–649.

3. Eugene Frankel, "J. B. Biot and the Mathematization of Experimental Physics in Napoleonic France," in *Historical Studies in the Physical Sciences*, ed. Russell McCormmach (Princeton, N.J.: Princeton University Press, 1977).

4. Charles Coulston Gillispie, ed., *Dictionary of Scientific Biography* (New York: Charles Scribner's Sons, 1981), p. 85.

5. For a discussion of the errors made by radar guns, see Nicole Weisensee Egan, "Takin' Aim at Radar Guns," *Philadelphia Daily News*, March 9, 2004.

6. Charles T. Clotfelter and Jacob L. Vigdor, "Retaking the SAT" (working paper SAN01-20, Terry Sanford Institute of Public Policy, Duke University, Durham, N.C., July 2001).

7. Eduardo Porter, "Jobs and Wages Increased Modestly Last Month," *New York Times*, September 2, 2006.

8. Gene Epstein on "Mathemagicians," *On the Media*, WNYC radio, broadcast August 25, 2006.

9. Legene Quesenberry et al., "Assessment of the Writing Component within a University General Education Program," November 1, 2000; http://wac.colostate.edu/aw/articles/quesenberry2000/quesenberry2000.pdf.

10. Kevin Saunders, "Report to the Iowa State University Steering Committee on the Assessment of ISU Comm-English 105 Course Essays," September 2004; www.iastate.edu/~isucomm/InYears/ISUcomm_essays.pdf (accessed 2005; site now discontinued).

11. University of Texas, Office of Admissions, "Inter-rater Reliability of Holistic Measures Used in the Freshman Admissions Process of the University of Texas at Austin," February 22, 2005; http://www.utexas.edu/student/admissions/research/Inter-raterReliability2005.pdf.

12. Emily J. Shaw and Glenn B. Milewski, "Consistency and Reliability in the Individualized Review of College Applicants," College Board, Office of Research and Development, *Research Notes* RN-20 (October 2004): 3; http://www.collegeboard.com/research/pdf/RN-20.pdf.

13. Gary Rivlin, "In Vino Veritas," *New York Times*, August 13, 2006.

14. William James, *The Principles of Psychology* (New York: Henry Holt, 1890), p. 509.

15. Robert Frank and Jennifer Byram, "Taste-Smell Interactions Are Tastant and Odorant Dependent," *Chemical Senses* 13 (1988): 445–55.

16. A. Rapp, "Natural Flavours of Wine: Correlation between Instrumental Analysis and Sensory Perception," *Fresenius' Journal of Analytic Chemistry* 337, no. 7 (January 1990): 777–85.

17. D. Laing and W. Francis, "The Capacity of Humans to Identify Odors in Mixtures," *Physiology and Behavior* 46, no. 5 (November 1989): 809–14; and D. Laing et al., "The Limited Capacity of Humans to Identify the Components of Taste Mixtures and Taste-Odour Mixtures," *Perception* 31, no. 5 (2002): 617–35.

18. For the rosé study, see Rose M. Pangborn, Harold W. Berg, and Brenda Hansen, "The Influence of Color on Discrimination of Sweetness in Dry Table-Wine," *American Journal of Psychology* 76, no. 3 (September 1963): 492–95. For the anthocyanin study, see G. Morrot, F. Brochet, and D. Dubourdieu, "The Color of Odors," *Brain and Language* 79, no. 2 (November 2001): 309–20.

19. Hilke Plassman, John O'Doherty, Baba Shia, and Antonio Rongel, "Marketing Actions Can Modulate Neural Representations of Experienced Pleasantness," *Proceedings of the National Academy of Sciences*, January 14, 2008; http://www.pnas.org.

20. M. E. Woolfolk, W. Castellan, and C. Brooks, "Pepsi versus Coke: Labels, Not Tastes, Prevail," *Psychological Reports* 52 (1983): 185–86.

21. M. Bende and S. Nordin, "Perceptual Learning in Olfaction: Professional Wine Tasters Versus Controls," *Physiology and Behavior* 62, no. 5 (November 1997): 1065–70.

22. Gregg E. A. Solomon, "Psychology of Novice and Expert Wine Talk," *American Journal of Psychology* 103, no. 4 (Winter 1990): 495–517.

23. Rivlin, "In Vino Veritas."

24. Ibid.

25. Hal Stern, "On the Probability of Winning a Football Game," *American Statistician* 45, no. 3 (August 1991): 179–82.

26. The graph is from Index Funds Advisors, "Index Funds.com: Take the Risk Capacity Survey," http://www.indexfunds3.com/step3page2.php, where it is credited to Walter Good and Roy Hermansen, *Index Your Way to Investment Success* (New York: New York Institute of Finance, 1997). The performance of 300 mutual fund managers was tabulated for ten years (1987–1996), based on the Morningstar Principia database.

27. Polling Report, "President Bush—Overall Job Rating," http://pollingreport.com/BushJob.htm.

28. "Poll: Bush Apparently Gets Modest Bounce," CNN, September 8, 2004, http://www.cnn.com/2004/ALLPOLITICS/09/06/presidential.poll/index.html.

29. "Harold von Braunhut," *Telegraph*, December 23, 2003; http://www.telegraph.co.uk/news/main.jhtml?xml=/news/2003/12/24/db2403.xml.

30. James J. Fogarty, "Why Is Expert Opinion on Wine Valueless?" (discussion paper 02.17, Department of Economics, University of Western Australia, Perth, 2001).

31. Stigler, *The History of Statistics*, p. 143.

Chapter 8: The Order in Chaos

1. Holland, *What Are the Chances?* p. 51.

2. This is only an approximation, based on more recent American statistics. See U.S. Social Security Administration, "Actuarial Publications: Period Life Table." The most recent table is available at http://www.ssa.gov/OACT/STATS/table4c6.html.

3. Immanuel Kant, quoted in Theodore Porter, *The Rise of Statistical Thinking: 1820–1900* (Princeton, N.J.: Princeton University Press, 1988), p. 51.

4. U.S. Department of Transportation, Federal Highway Administration, "Licensed Drivers, Vehicle Registrations and Resident Population," http://www.fhwa.dot.gov/policy/ohim/hs03/htm/dlchrt.htm.

5. U.S. Department of Transportation, Research and Innovative Technology Administration, Bureau of Transportation Statistics, "Motor Vehicle Safety Data," http://www.bts.gov/publications/national_transportation_statistics/2002/html/table_02_17.html.

6. "The Domesday Book," *History Magazine*, October/November 2001.

7. For Graunt's story, see Hacking, *The Emergence of Probability*, pp. 103–9; David, *Gods, Games and Gambling*, pp. 98–109; and Newman, *The World of Mathematics*, 3:1416–18.

8. Hacking, *The Emergence of Probability*, p. 102.

9. Theodore Porter, *The Rise of Statistical Thinking*, p. 19.

10. For Graunt's original table, see Hacking, *The Emergence of Probability*, p. 108. For the current data, see World Health Organization, "Life Tables for WHO Member States," http://www.who.int/whosis/database/life_tables/life_tables.cfm. The figures quoted were taken from abridged tables and rounded.

11. Ian Hacking, *The Taming of Chance* (Cambridge: Cambridge University Press, 1990), p. vii.

12. H. A. David, "First (?) Occurrence of Common Terms in Statistics and Probability," in *Annotated Readings in the History of Statistics*, ed. H. A. David and A.W.F. Edwards (New York: Springer, 2001), appendix B and pp. 219–28.

13. Noah Webster, *American Dictionary of the English Language* (1828; facsimile of the 1st ed., Chesapeake, Va.: Foundation for American Christian Education, 1967).

14. The material on Quételet is drawn mainly from Stigler, *The History of Statistics*, pp. 161–220; Stephen Stigler, *Statistics on the Table: The History of Statistical Concepts and Methods* (Cambridge, Mass.: Harvard University Press, 1999), pp. 51–66; and Theodore Porter, *The Rise of Statistical Thinking*, pp. 100–9.

15. Louis Menand, *The Metaphysical Club* (New York: Farrar, Straus & Giroux, 2001), p. 187.

16. Holland, *What Are the Chances?* pp. 41–42.

17. David Yermack, "Good Timing: CEO Stock Option Awards and Company News Announcements," *Journal of Finance* 52, no. 2 (June 1997): 449–76; and Erik Lie, "On the Timing of CEO Stock Option Awards," *Management Sci-*

ence 51, no. 5 (May 2005): 802–12. See also Charles Forelle and James Bandler, "The Perfect Payday—Some CEOs Reap Millions by Landing Stock Options When They Are Most Valuable: Luck—or Something Else?" *Wall Street Journal,* March 18, 2006.

18. Justin Wolfers, "Point Shaving: Corruption in NCAA Basketball," *American Economic Review* 96, no. 2 (May 2006): 279–83.

19. Stern, "On the Probability of Winning a Football Game."

20. David Leonhardt, "Sad Suspicions about Scores in Basketball," *New York Times,* March 8, 2006.

21. Richard C. Hollinger et al., *National Retail Security Survey: Final Report* (Gainesville: Security Research Project, Department of Sociology and Center for Studies in Criminal Law, University of Florida, 2002–2006).

22. Adolphe Quételet, quoted in Theodore Porter, *The Rise of Statistical Thinking,* p. 54.

23. Quételet, quoted in Menand, *The Metaphysical Club,* p. 187.

24. Jeffrey Kluger, "Why We Worry about the Things We Shouldn't . . . and Ignore the Things We Should," *Time,* December 4, 2006, pp. 65–71.

25. Gerd Gigerenzer, *Empire of Chance: How Probability Changed Science and Everyday Life* (Cambridge: Cambridge University Press, 1989), p. 129.

26. Menand, *The Metaphysical Club,* p. 193.

27. De Vany, *Hollywood Economics;* see part IV, "A Business of Extremes."

28. See Derek William Forrest, *Francis Galton: The Life and Work of a Victorian Genius* (New York: Taplinger, 1974); Jeffrey M. Stanton, "Galton, Pearson, and the Peas: A Brief History of Linear Regression for Statistics Instructors," *Journal of Statistics Education* 9, no. 3 (2001); and Theodore Porter, *The Rise of Statistical Thinking,* pp. 129–46.

29. Francis Galton, quoted in Theodore Porter, *The Rise of Statistical Thinking,* p. 130.

30. Peter Doskoch, "The Winning Edge," *Psychology Today,* November/December 2005, pp. 44–52.

31. Deborah J. Bennett, *Randomness* (Cambridge, Mass.: Harvard University Press, 1998), p. 123.

32. Abraham Pais, *The Science and Life of Albert Einstein* (London: Oxford University Press, 1982), p. 17; see also the discussion on p. 89.

33. On Brown and the history of Brownian motion, see D. J. Mabberley, *Jupiter Botanicus: Robert Brown of the British Museum* (Braunschweig, Germany, and London: Verlag von J. Cramer / Natural History Museum, 1985); Brian J. Ford, "Brownian Movement in Clarkia Pollen: A Reprise of the First Observations," *Microscope* 40, no. 4 (1992): 235–41; and Stephen Brush, "A History of Random Processes. I. Brownian Movement from Brown to Perrin," *Archive for History of Exact Sciences* 5, no. 34 (1968).

34. Pais, *Albert Einstein,* pp. 88–100.

35. Albert Einstein, quoted in Ronald William Clark, *Einstein: The Life and Times* (New York: HarperCollins, 1984), p. 77.

Chapter 9: Illusions of Patterns and Patterns of Illusion

1. See Arthur Conan Doyle, *The History of Spiritualism* (New York: G. H. Doran, 1926); and R. L. Moore, *In Search of White Crows: Spiritualism, Parapsychology, and American Culture* (London: Oxford University Press, 1977).

2. Ray Hyman, "Parapsychological Research: A Tutorial Review and Critical Appraisal," *Proceedings of the IEEE* 74, no. 6 (June 1986): 823–49.

3. Michael Faraday, "Experimental Investigation of Table-Moving," *Athenaeum*, July 2, 1853, pp. 801–3.

4. Michael Faraday, quoted in Hyman, "Parapsychological Research," p. 826.

5. Faraday, quoted ibid.

6. See Frank H. Durgin, "The Tinkerbell Effect: Motion Perception and Illusion," *Journal of Consciousness Studies* 9, nos. 5–6 (May–June 2002): 88–101.

7. Christof Koch, *The Quest for Consciousness: A Neurobiological Approach* (Englewood, Colo.: Roberts, 2004), pp. 51–54.

8. The study was D. O. Clegg et al., "Glucosamine, Chondroitin Sulfate, and the Two in Combination for Painful Knee Osteoarthritis," *New England Journal of Medicine* 354, no. 8 (February 2006): 795–808. The interview was "Slate's Medical Examiner: Doubts on Supplements," *Day to Day*, NPR broadcast, March 13, 2006.

9. See Paul Slovic, Howard Kunreuther, and Gilbert F. White, "Decision Processes, Rationality, and Adjustment to Natural Hazards," in *Natural Hazards: Local, National, and Global*, ed. G. F. White (London: Oxford University Press, 1974); see also Willem A. Wagenaar, "Generation of Random Sequences by Human Subjects: A Critical Survey of Literature," *Psychological Bulletin* 77, no. 1 (January 1972): 65–72.

10. See Hastie and Dawes, *Rational Choice in an Uncertain World*, pp. 19–23.

11. George Spencer-Brown, *Probability and Scientific Inference* (London: Longmans, Green, 1957), pp. 55–56. Actually, 10 is a gross underestimate.

12. Janet Maslin, "His Heart Belongs to (Adorable) iPod," *New York Times*, October 19, 2006.

13. Hans Reichenbach, *The Theory of Probability*, trans. E. Hutton and M. Reichenbach (Berkeley: University of California Press, 1934).

14. The classic text expounding this point of view is Burton G. Malkiel, *A Random Walk Down Wall Street*, now completely revised in an updated 8th ed. (New York: W. W. Norton, 2003).

15. John R. Nofsinger, *Investment Blunders of the Rich and Famous—and What You Can Learn from Them* (Upper Saddle River, N.J.: Prentice Hall, Financial Times, 2002), p. 62.

16. Hemang Desai and Prem C. Jain, "An Analysis of the Recommendations of the 'Superstar' Money Managers at *Barron's* Annual Roundtable," *Journal of Finance* 50, no. 4 (September 1995): 1257–73.

17. Jess Beltz and Robert Jennings, "*Wall $treet Week with Louis Rukeyser*'s Recommendations: Trading Activity and Performance," *Review of Financial*

Economics 6, no. 1 (1997): 15–27; and Robert A. Pari, "*Wall $treet Week* Recommendations: Yes or No?" *Journal of Portfolio Management* 14, no. 1 (1987): 74–76.

18. Andrew Metrick, "Performance Evaluation with Transactions Data: The Stock Selection of Investment Newsletters, *Journal of Finance* 54, no. 5 (October 1999): 1743–75; and "The Equity Performance of Investment Newsletters" (discussion paper no. 1805, Harvard Institute of Economic Research, Cambridge, Mass., November 1997).

19. James J. Choi, David Laibson, and Brigitte Madrian, "Why Does the Law of One Price Fail? An Experiment on Index Mutual Funds" (working paper no. W12261, National Bureau of Economic Research, Cambridge, Mass., May 4, 2006).

20. Leonard Koppett, "Carrying Statistics to Extremes," *Sporting News*, February 11, 1978.

21. By some definitions, Koppett's system would be judged to have failed in 1970; by others, to have passed. See CHANCE News 13.04, April 18, 2004–June 7, 2004, http://www.dartmouth.edu/~chance/chance_news/recent_news/chance_news_13.04.html.

22. As touted on the Legg Mason Capital Management Web site, http://www.leggmasoncapmgmt.com/awards.htm.

23. Lisa Gibbs, "Miller: He Did It Again," CNNMoney, January 11, 2004, http://money.cnn.com/2004/01/07/funds/ultimateguide_billmiller_0204.

24. Thomas R. Gilovich, Robert Vallone, and Amos Tversky, "The Hot Hand in Basketball: On the Misperception of Random Sequences," *Cognitive Psychology* 17, no. 3 (July 1985): 295–314.

25. Purcell's research is discussed in Gould, "The Streak of Streaks."

26. Mark Hulbert, "Not All Stocks Are Created Equal," www.MarketWatch.com, January 18, 2005, accessed March 2005 (site now discontinued).

27. Kunal Kapoor, "A Look at Who's Chasing Bill Miller's Streak," Morningstar, December 30, 2004, http://www.morningstar.com.

28. Michael Mauboussin and Kristen Bartholdson, "On Streaks: Perception, Probability, and Skill," *Consilient Observer* (Credit Suisse–First Boston) 2, no. 8 (April 22, 2003).

29. Merton Miller on "Trillion Dollar Bet," *NOVA*, PBS broadcast, February 8, 2000.

30. R. D. Clarke, "An Application of the Poisson Distribution," *Journal of the Institute of Actuaries* 72 (1946): 48.

31. Atul Gawande, "The Cancer Cluster Myth," *The New Yorker*, February 28, 1998, pp. 34–37.

32. Ibid.

33. Bruno Bettelheim, "Individual and Mass Behavior in Extreme Situations," *Journal of Abnormal and Social Psychology* 38 (1943): 417–52.

34. Curt P. Richter, "On the Phenomenon of Sudden Death in Animals and Man," *Psychosomatic Medicine* 19 (1957): 191–98.

35. E. Stotland and A. Blumenthal, "The Reduction of Anxiety as a Result of the Expectation of Making a Choice," *Canadian Review of Psychology* 18 (1964): 139–45.

36. Ellen Langer and Judith Rodin, "The Effects of Choice and Enhanced Personal Responsibility for the Aged: A Field Experiment in an Institutional Setting," *Journal of Personality and Social Psychology* 34, no. 2 (1976): 191–98.

37. Ellen Langer and Judith Rodin, "Long-Term Effects of a Control-Relevant Intervention with the Institutionalized Aged," *Journal of Personality and Social Psychology* 35, no. 12 (1977): 897–902.

38. L. B. Alloy and L. Y. Abramson, "Judgment of Contingency in Depressed and Nondepressed Students: Sadder but Wiser?" *Journal of Experimental Psychology: General* 108, no. 4 (December 1979): 441–85.

39. Durgin, "The Tinkerbell Effect."

40. Ellen Langer, "The Illusion of Control," *Journal of Personality and Social Psychology* 32, no. 2 (1975): 311–28.

41. Ellen Langer and Jane Roth, "Heads I Win, Tails It's Chance: The Illusion of Control as a Function of Outcomes in a Purely Chance Task," *Journal of Personality and Social Psychology* 32, no. 6 (1975): 951–55.

42. Langer, "The Illusion of Control."

43. Ibid., p. 311.

44. Raymond Fisman, Rakesh Khurana, and Matthew Rhodes-Kropf, "Governance and CEO Turnover: Do Something or Do the Right Thing?" (working paper no. 05-066, Harvard Business School, Cambridge, Mass., April 2005).

45. P. C. Wason, "Reasoning about a Rule," *Quarterly Journal of Experimental Psychology* 20 (1968): 273–81.

46. Francis Bacon, *Novum Organon,* trans. by P. Urbach and J. Gibson (Chicago: Open Court, 1994), p. 57 (originally published in 1620).

47. Charles G. Lord, Lee Ross, and Mark Lepper, "Biased Assimilation and Attitude Polarization: The Effects of Prior Theories on Subsequently Considered Evidence," *Journal of Personality and Social Psychology* 37, no. 11 (1979): 2098–109.

48. Matthew Rabin, "Psychology and Economics" (white paper, University of California, Berkeley, September 28, 1996).

49. E. C. Webster, *Decision Making in the Employment Interview* (Montreal: Industrial Relations Centre, McGill University, 1964).

50. Beth E. Haverkamp, "Confirmatory Bias in Hypothesis Testing for Client-Identified and Counselor Self-generated Hypotheses," *Journal of Counseling Psychology* 40, no. 3 (July 1993): 303–15.

51. David L. Hamilton and Terrence L. Rose, "Illusory Correlation and the Maintenance of Stereotypic Beliefs," *Journal of Personality and Social Psychology* 39, no. 5 (1980): 832–45; Galen V. Bodenhausen and Robert S. Wyer, "Effects of Stereotypes on Decision Making and Information-Processing Strategies," *Journal of Personality and Social Psychology* 48, no. 2 (1985): 267–82; and C. Stangor and D. N. Ruble, "Strength of Expectancies and Memory for Social Information:

What We Remember Depends on How Much We Know," *Journal of Experimental Social Psychology* 25, no. 1 (1989): 18–35.

Chapter 10: The Drunkard's Walk

1. Pierre-Simon de Laplace, quoted in Stigler, *Statistics on the Table*, p. 56.

2. James Gleick, *Chaos: Making a New Science* (New York: Penguin, 1987); see chap. 1.

3. Max Born, *Natural Philosophy of Cause and Chance* (Oxford: Clarendon Press, 1948), p. 47. Born was referring to nature in general and quantum theory in particular.

4. The Pearl Harbor analysis is from Roberta Wohlstetter, *Pearl Harbor: Warning and Decision* (Palo Alto, Calif.: Stanford University Press, 1962).

5. Richard Henry Tawney, *The Agrarian Problem in the Sixteenth Century* (1912; repr., New York: Burt Franklin, 1961).

6. Wohlstetter, *Pearl Harbor*, p. 387.

7. The description of the events at Three Mile Island is from Charles Perrow, *Normal Accidents: Living with High-Risk Technologies* (Princeton, N.J.: Princeton University Press, 1999); and U.S. Nuclear Regulatory Commission, "Fact Sheet on the Three Mile Island Accident," http://www.nrc.gov/reading-rm/doc-collections/fact-sheets/3mile-isle.html.

8. Perrow, *Normal Accidents*.

9. W. Brian Arthur, "Positive Feedbacks in the Economy," *Scientific American*, February 1990, pp. 92–99.

10. Matthew J. Salganik, Peter Sheridan Dodds, and Duncan J. Watts, "Experimental Study of Inequality and Unpredictability in an Artificial Cultural Market," *Science* 311 (February 10, 2006); and Duncan J. Watts, "Is Justin Timberlake a Product of Cumulative Advantage?" *New York Times Magazine*, April 15, 2007.

11. Mlodinow, "Meet Hollywood's Latest Genius."

12. John Steele Gordon and Michael Maiello, "Pioneers Die Broke," *Forbes*, December 23, 2002; and "The Man Who Could Have Been Bill Gates," *BusinessWeek*, October 25, 2004.

13. Floyd Norris, "Trump Deal Fails, and Shares Fall Again," *New York Times*, July 6, 2007.

14. Melvin J. Lerner and Leo Montada, "An Overview: Advances in Belief in a Just World Theory and Methods," in *Responses to Victimizations and Belief in a Just World*, ed. Leo Montada and Melvin J. Lerner (New York: Plenum Press, 1998), pp. 1–7.

15. Melvin J. Lerner, "Evaluation of Performance as a Function of Performer's Reward and Attractiveness," *Journal of Personality and Social Psychology* 1 (1965): 355–60.

16. Melvin J. Lerner and C. H. Simmons, "Observer's Reactions to the 'Innocent Victim': Compassion or Rejection?" *Journal of Personality and Social Psychology* 4 (1966): 203–10.

17. Lerner, "Evaluation of Performance as a Function of Performer's Reward and Attractiveness."

18. Wendy Jean Harrod, "Expectations from Unequal Rewards," *Social Psychology Quarterly* 43, no. 1 (March 1980): 126–30; Penni A. Stewart and James C. Moore Jr., "Wage Disparities and Performance Expectations," *Social Psychology Quarterly* 55, no. 1 (March 1992): 78–85; and Karen S. Cook, "Expectations, Evaluations and Equity," *American Sociological Review* 40, no. 3 (June 1975): 372–88.

19. Lerner and Simmons, "Observer's Reactions to the 'Innocent Victim.' "

20. David L. Rosenhan, "On Being Sane in Insane Places," *Science* 179 (January 19, 1973): 237–55.

21. Elisha Y. Babad, "Some Correlates of Teachers' Expectancy Bias," *American Educational Research Journal* 22, no. 2 (Summer 1985): 175–83.

22. Eric Asimov, "Spirits of the Times: A Humble Old Label Ices Its Rivals," *New York Times*, January 26, 2005.

23. Jonathan Calvert and Will Iredale, "Publishers Toss Booker Winners into the Reject Pile," *London Sunday Times*, January 1, 2006.

24. Peter Doskoch, "The Winning Edge," *Psychology Today*, November/December 2005, p. 44.

25. Rosenhan, "On Being Sane in Insane Places," p. 243.

INDEX

Page numbers from 223 refer to endnotes;
page numbers in *italics* refer to graphs or tables.

239

Index

Index

Hannibal, 31
Harry Potter (Rowling), 10, 125
Harvard Institute of Economic Research, 176
Harvard University, 104, 186, 188
Hebrew University, 7, 82
height, 163
Hereditary Genius (Galton), 161
heredity, 161–63
Herodotus, 27
heuristics, 174
Hill, Ted, 83
Hillerman, Tony, 10
Hindus, 49
history, 201
History of Civilization in England (Buckle), 159, 165
Hitler, Adolf, 4
HIV tests, 115–16, 229
Hoffman, Dustin, 12
Holiday (Middleton), 215
Hollywood, *see* movie industry
homeless persons, 28–29
Homer, 27
homme moyen, l' (average man), 158, 159
hot and cold streaks, 14, 101, 175, 177–82
hot-hand fallacy, 178–83
"Hot Hand in Basketball, The" (Tversky), 178–79
Huygens, Christiaan, 89–90

IAAF (International Association of Athletics Federations), 117
IBM, 208–9, 217
imagination, perception and, 170–71
In a Free State (Naipaul), 215
industry, 111, 143
infinity, 93, 94
Inquisition, 41, 58, 65
insanity, 213–14
insurance:
    car, 111–12
    life, 109–10, 114–15, 152

Internal Revenue Service (IRS), 84
International Association of Athletics Federations (IAAF), 117
Internet, 205
intuition:
    probabilistic, 22–26
    as unsuited for interpreting randomness, ix–xi, 3–20, 45, 99, 174
inversion error (prosecutor's fallacy), 118–21
Iowa State University, 130
iPod, 175
IQ ratings, 136
Ireland, 151
irrationality, 5, 194
Ishtar, 12
isomorphism, 52
Israeli air force, 7–8

Jagger, Joseph, 85–88, 90, 164
James, William, 131, 135
Japan, 195–96
Jarvik, Robert, 43
Jennings, Ken, 177
Jeopardy, 177
Jerry Maguire, 16
Jia Xian, 72
job interviews, 191
Jobs, Steve, 175
Jones, Paula, 25
Journal of the Institute of Actuaries, 183
Justinian, Emperor, 32

Kahneman, Daniel, 28, 99
    heuristics and, 174
    Nobel Prize awarded to, 7
    randomness studies of Tversky and, 7–9, 22–25, 194
Kant, Immanuel, 93, 147
Kerrich, John, 95, 101
Kildall, Gary, 208–9
King, Stephen, 205, 215–16, 217
Koppett, Leonard, 177, 181, 235

Index

SIDS (sudden infant death syndrome), 118–19
significance testing, 171–73
Simpson, Nicole Brown, 119–20
Simpson, O. J., 119–21
Singer, Isaac Bashevis, 10
Skinner, B. F., 105
Slaney, Mary Decker, 117–18
small numbers, law of, 99–101
SmartMoney, 178
S-matrix theory, 216
smell, 132
Smith, Adam, 109
"social physics," 158–60
Social Security Administration, 113
sociology, 204–5
Socrates, 28
Sophocles, 27
Spencer-Brown, George, 174–75, 176
spiritualism, 169–70
split brain, 6
Sporting News, 177
sports:
  drug testing in, 117–18
  fraud in, 156–58
  hot-hand fallacy and, 178–79
  inferior team's winning of a series in, 70–71
  misplaced emphasis on coaches in, x, 188
  randomness of extraordinary achievements in, 16–20
  trading cards lottery, 187
  see also specific sports
standard deviation, 138–43, 157
  see also sample standard deviation
Star Wars, 12–14
statistical analysis, 171–73
statistical physics, 165–68
statistical significance, 71, 98
statistics, x
  biological applications of, 160–65
  chi-square test and, 164–65
  coefficient of correlation and, 163

on crime, 154–55, 158, 159
derivation of term, 153
forensic economics and, 156
founders of, 150–51
l'homme moyen and, 158
mathematical, 129
mathematicians' attitude toward, 150
on mortality, 155
probability and, 122–23
real-world applicability of, 129
significance testing and, 171–73
social data and, 148–68
"social physics" and, 158–60
Statistik, 153
Stigler, Stephen, 102
stock market, 176–77, 198
  see also mutual funds
stock options, 156
string theory, 216
success:
  in cultural marketing, 204–5
  normal accident theory and, 204–10
  outward signs of, 209–11
  persistence and, 10–11, 161–62, 216
  randomness and, 11–20, 210–17
  random rewards and, 209–10
  in random trials, 96
  short-term, 70–71, 99–101
  Watson's maxim on, 217
sudden infant death syndrome (SIDS), 118–19
suicide, 155
sum (third) law of probability, 35
Super Bowl, 177
supermarket lines, 29
superstitions, 46, 60–61

table moving, 169–70
Tartaglia, Niccolò, 58
taste, 131–32
Tawney, Richard Henry, 201

250

A NOTE ON THE TYPE

THIS BOOK was set in Electra, a typeface designed by William Addison Dwiggins (1880–1956) for the Mergenthaler Linotype Company and first made available in 1935. Electra cannot be classified as either "modern" or "old style." It is not based on any historical model, and hence does not echo any particular period or style of type design. It avoids the extreme contrast between thick and thin elements that marks most modern faces, and it is without eccentricities that catch the eye and interfere with reading. In general, Electra is a simple, readable typeface that attempts to give a feeling of fluidity, power, and speed.

W. A. Dwiggins was born in Martinsville, Ohio, and studied art in Chicago. In the late 1920s he moved to Hingham, Massachusetts, where he built a solid reputation as a designer of advertisements and as a calligrapher. He began an association with the Mergenthaler Linotype Company in 1929 and over the next twenty-seven years designed a number of book types of which Metro, Electra, and Caledonia have been used widely. In 1930 Dwiggins became interested in marionettes, and through the years he made many important contributions to the art of puppetry and the design of marionettes.

Composed by North Market Street Graphics,
Lancaster, Pennsylvania
Printed and bound by Berryville Graphics,
Berryville, Virginia
Designed by Virginia Tan